Springer Theses

Recognizing Outstanding Ph.D. Research

Aims and Scope

The series "Springer Theses" brings together a selection of the very best Ph.D. theses from around the world and across the physical sciences. Nominated and endorsed by two recognized specialists, each published volume has been selected for its scientific excellence and the high impact of its contents for the pertinent field of research. For greater accessibility to non-specialists, the published versions include an extended introduction, as well as a foreword by the student's supervisor explaining the special relevance of the work for the field. As a whole, the series will provide a valuable resource both for newcomers to the research fields described, and for other scientists seeking detailed background information on special questions. Finally, it provides an accredited documentation of the valuable contributions made by today's younger generation of scientists.

Theses may be nominated for publication in this series by heads of department at internationally leading universities or institutes and should fulfill all of the following criteria

- They must be written in good English.
- The topic should fall within the confines of Chemistry, Physics, Earth Sciences, Engineering and related interdisciplinary fields such as Materials, Nanoscience, Chemical Engineering, Complex Systems and Biophysics.
- The work reported in the thesis must represent a significant scientific advance.
- If the thesis includes previously published material, permission to reproduce this must be gained from the respective copyright holder (a maximum 30% of the thesis should be a verbatim reproduction from the author's previous publications).
- They must have been examined and passed during the 12 months prior to nomination.
- Each thesis should include a foreword by the supervisor outlining the significance of its content.
- The theses should have a clearly defined structure including an introduction accessible to new PhD students and scientists not expert in the relevant field.

Indexed by zbMATH.

Van Hiep Nguyen

Bioinspired Multifunctional Nanomaterials for Ionic Artificial Muscles

Doctoral Thesis accepted by
Korea Advanced Institute of Science and Technology (KAIST), Daejeon, Korea

Springer

Author
Dr. Van Hiep Nguyen
Korea Advanced Institute of Science
and Technology (KAIST)
Daejeon, Korea (Republic of)

Supervisor
Prof. Dr. Il-Kwon Oh
Department of Mechanical Engineering
Korea Advanced Institute of Science
and Technology (KAIST)
Daejeon, Korea (Republic of)

ISSN 2190-5053 ISSN 2190-5061 (electronic)
Springer Theses
ISBN 978-3-031-78812-3 ISBN 978-3-031-78813-0 (eBook)
https://doi.org/10.1007/978-3-031-78813-0

This Springer imprint is published by the registered company Springer Nature Switzerland AG
The registered company address is: Gewerbestrasse 11, 6330 Cham, Switzerland

To my family

Supervisor's Foreword

Ionic electroactive polymer actuators, which consist of two electrodes enclosing a polymer electrolyte membrane, undergo deformation as a result of ion migration within the membrane when subjected to an electric field generated by the electrodes. While these actuators are regarded as promising candidates for artificial muscles in soft robotics, their practical implementation poses significant challenges due to the complex requirements for multifunctional materials and the necessity for interdisciplinary research. Addressing these challenges is the central focus of Van Hiep Nguyen's doctoral research.

At the outset of his Ph.D. studies on ionic actuators, Mr. Nguyen identified key material properties that exhibit inverse trends, a concept thoroughly explored in Chap. 1. His observations of naturally occurring multifunctional structures inspired innovative strategies for developing materials with these inverse-trend properties. Leveraging his strong foundation in chemistry and materials science, he devised synthesis procedures for creating such materials, which included two high-performance electrolyte membranes with superior ionic conductivity and two capacitive electrode materials with enhanced electrical conductivity. The resulting multifunctional materials led to significant improvements in actuation performance, facilitating several novel demonstrations of soft robotics, which are comprehensively detailed in Chaps. 2–5.

This thesis successfully bridges the gap between materials science, ionic actuators, and soft robotics, pushing the boundaries of ionic actuators toward practical engineering applications. The methodologies and materials developed in this work provide valuable insights for researchers not only within the field of ionic actuators but also in adjacent disciplines, particularly those focused on energy storage and conversion, where multifunctional materials are critical.

Daejeon, Korea (Republic of)
September 2024

Prof. Dr. Il-Kwon Oh

Abstract

Ionic electroactive polymer actuators, a type of artificial muscles featuring a polymer electrolyte membrane sandwiched between two electrodes, hold significant potential for soft robotic applications. However, their practical use demands enhanced performance, including faster response times, greater bending deformation, stronger blocking forces, and improved durability. Achieving these advancements requires multifunctional nanomaterials that exhibit high ionic conductivity and robust mechanical properties for electrolytes, alongside rapid electrical conductivity and high capacitance for electrodes. Unfortunately, these properties are often in opposition and, hence, rarely found in conventional materials.

Inspired by nature, which integrates such properties into multifunctional platforms, this dissertation investigates four innovative approaches to developing nanomaterials that satisfy these demanding requirements. The first two approaches focus on fabricating robust, micro-thick electrolyte membranes with high ionic conductivity. This is accomplished through a block isomer inspired by cell membranes and a hemostasis-inspired method to control the molecular configuration of a Nafion isomer. The latter two approaches aim to develop electrodes with rapid conductivity and high capacitance. This is achieved using a molybdenum disulfide-graphene nanocomposite inspired by calcium metabolism and a covalent triazine framework inspired by bone structure. As a result, these materials show significant improvements in actuation performance, enabling applications such as soft fingers, inchworms, dynamic tensegrity structures, and dragonflies.

The multidisciplinary research presented in this dissertation advances the fields of materials science, ionic actuators, soft robotics, bioinspiration, and biomimetics. The developed materials hold potential for applications in batteries, capacitors, and fuel cells. Additionally, the nature-inspired methodologies outlined here could be adapted to develop new materials with similarly challenging properties for a wide range of applications.

Related Publications

V. H. Nguyen, J. Kim, R. Tabassian, M. Kotal, K. Jun, J.-H. Oh, J.-M. Son, M. T. Manzoor, K. J. Kim, I.-K. Oh*, "*Electroactive artificial muscles based on functionally antagonistic core-shell polymer electrolyte derived from PS-b-PSS block copolymer*", Advanced Science 1801196, 2019.

V. H. Nguyen, S. Oh, M. Mahato, R. Tabassian, H. Yoo, S.-G. Lee, M. Garai, K. J. Kim, I.-K. Oh*, "*Functionally Antagonistic Polyelectrolyte for Electro-Ionic Soft Actuator*", Nature Communications 435, 2024.

M. T. Manzoor, V. H. Nguyen, S. Umrao, J. Kim, R. Tabassian, J. E. Kim, I.-K. Oh, "*Mutually exclusive p-type and n-type hybrid electrode of* MoS_2 *and graphene for artificial soft touch fingers*", Advanced Functional Materials 1905454, 2019, M. T. M. and V. H. N. equal contribution.

M. Garai, V. H. Nguyen, M. Mahato, S. Oh, D. Saatchi, H. Yoo, S. S. Ali, D. V. Lam, A. K. Taseer, I.-K. Oh*, "*Multifunctional and electronically conjugated triazine framework for superior electro-ionic artificial muscles*", Advanced Functional Materials, 2406603, 2024, M.G. and V. H. N. equal contribution.

Acknowledgments

First, I would like to express my deep gratitude to Prof. Il-Kwon Oh, my academic advisor, for his unwavering support, which has been pivotal to my success in the research presented in this doctoral dissertation. His encouragement has driven me to explore my abilities, identify my research topics and directions, and overcome challenges throughout my studies. His trust has given me the confidence to pursue intellectual and professional growth, which will be crucial for my future career.

Second, I extend my heartfelt thanks to my family and friends for their consistent support and unconditional love, which have significantly contributed to my success. This achievement is dedicated to them.

Third, I acknowledge all the laboratory members for their kind assistance and collaboration. Together, we have cultivated a professional research environment that fosters our scientific passion.

Lastly, I would like to thank the committee members for evaluating this dissertation and offering valuable advice, the professors who taught me in their courses, and the staff who assisted me during my time at KAIST.

Contents

Chapter 1
Introduction to Ionic Artificial Muscles

1.1 Fundamentals of Ionic Polymer Electroactive Artificial Muscles

The natural muscle, composed of living cells, responds to neural signals by contracting and converting the chemical energy stored in adenosine triphosphate into mechanical energy [1]. This biological process underlies all animal movement. Similarly, artificial muscles, or soft actuators, are made from compliant materials that respond to external stimuli, deforming and converting input energy into mechanical work [1–3]. Unlike natural muscles, the stimuli for artificial muscles often serve as their input energy. These stimuli include electrical [1, 4–10] and magnetic [11–13] fields, electromagnetic waves [14–16], pH [17–20], concentration of salt solutions [21], temperature [22–25], humidity [26, 27], solvents [28], hydraulics [29, 30], and pneumatics [31–33], etc. Specifically, electroactive artificial muscles use electrical field as both stimuli and input energy [3, 5, 9, 34–36]. This category includes ionic polymer actuators, which deform in response to ion currents within an electrolyte membrane positioned between two electrodes [9, 37–44].

1.1.1 Materials and Mechanism

Building on pioneering research into the electrically responsive polymers [38–40, 42, 45], the first ionic actuator was developed in 1994 [46]. Its structure, illustrated in Fig. 1.1a, consists of an electrolyte membrane coated with two electrodes [9, 37–42]. Since the first prototype, ionic actuators have advanced significantly due to innovations in both electrolyte and electrode materials. Common materials are summarized in Fig. 1.1b and Table 1.1.

The initial electrolyte was poly(vinyl alcohol) (PVA), followed by various innovative polymers [46]. Among these, Nafion and block copolymers are particularly

V. H. Nguyen, *Bioinspired Multifunctional Nanomaterials for Ionic Artificial Muscles*, Springer Theses, https://doi.org/10.1007/978-3-031-78813-0_1

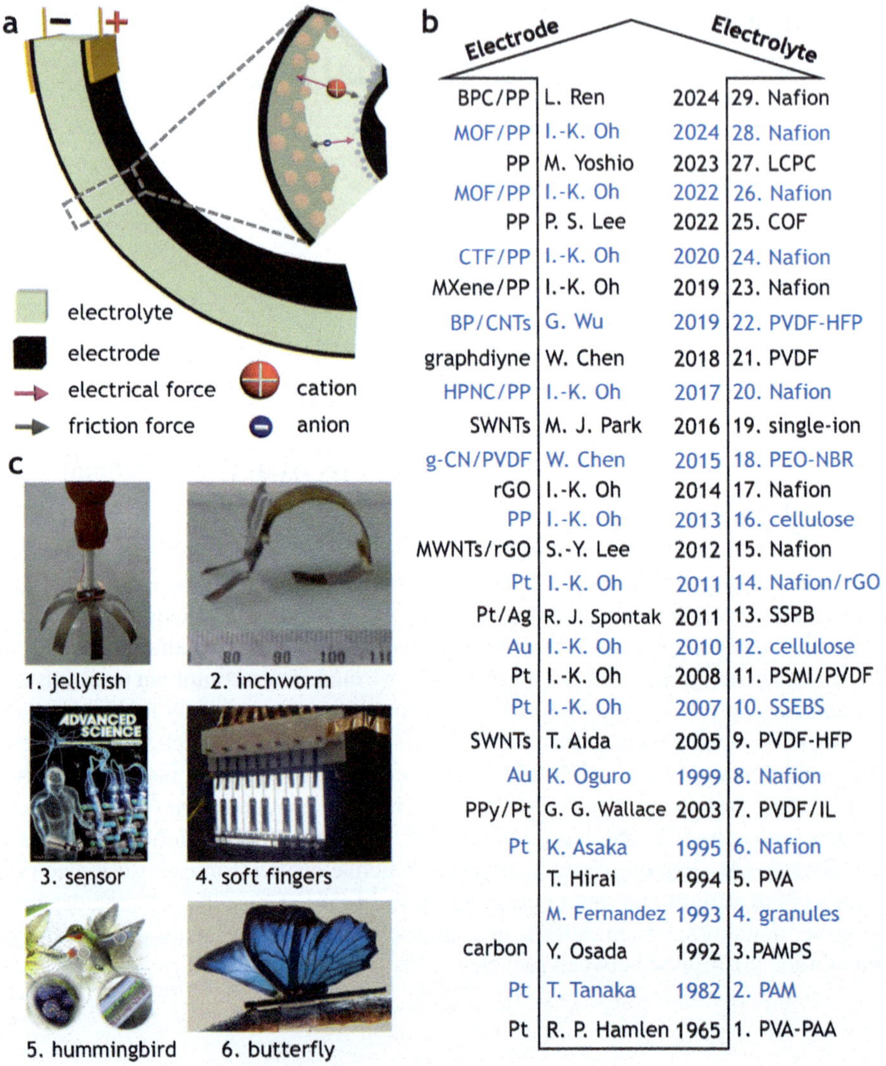

Fig. 1.1 Ionic artificial muscles: **a** materials, structure, and mechanism; **b** brief history, 1. PVA-PAA [40], 2. PAM [39], 3. PAMPS [38], 4. granules [42], 5. PVA [46], 6. Nafion [55], 7. PVDF/IL [56], 8. Nafion [57], 9. PVDF-HFP [58], 10. SSEBS [52], 11. PSMI/PVDF [59], 12. cellulose [60], 13. SSPB [53], 14. Nafion/rGO [61], 15. Naifon [62], 16. cellulose [63], 17. Nafion [64], 18. PEO-NBR [65], 19. single-ion [66], 20. Naifon [67], 21. PVDF [68], 22. PVDF-HFP [69], 23. Nafion [70], 24. Nafion [71], 25. COF [72], 26. Nafion [73], 27. LCPC [74], 28. Nafion [75], 29. Nafion [76]; **c** applications, 1. jellyfish [77] from Ref. [77]. Reprinted with permission from IOP Publishing, 2. inchworm [78] from Ref. [78]. Reprinted with permission from John Wiley and Sons, 3. self-sensor [79] from Ref. [79]. Reprinted with permission from John Wiley and Sons, 4. soft fingers [71] from Ref. [71]. Reprinted with permission from Springer Nature, 5. hummingbird [73] from Ref. [73]. Reprinted with permission from John Wiley and Sons, 6. butterfly [70] from Ref. [70]. Reprinted with permission from AAAS

Table 1.1 Typical electrolyte and electrode materials used in ionic artificial muscles

Electrolyte materials	Electrode materials	Blocking force, mN	Rise time, s	Strain, %
Nafion/ $EMImBF_4$ (EB) [76]	BPC/PP	–	–	0.54 (0.5 V, 0.1 Hz)
Nafion/EB [75]	Cobalt MOF/PP	47.5 (1.0V)	2.1	0.98 (1.0 V)
LCPC/LiTFSI [74]	PP	0.35 (2.0V)	–	0.63 (2.0 V, 0.1 Hz)
Nafion/EB [73]	Nickel MOF/PP	5.80 (2.0V)	1.5	0.72 (1.0 V, 0.1 Hz)
COF [72]	PP	1.20 (0.5V)	1.0	0.38 (0.5 V, 1 Hz)
Nafion/EB [71]	CTF/PP	2.79 (2.0V)	10.0	0.38 (0.5 V, 0.1 Hz)
Nafion/EB [70]	$Ti_3C_2T_x$ MXene/ PP	4.71 (1.0V)	1.0	1.37 (1.0 V, 0.1 Hz)
PVDF-HFP/EB [69]	BP/CNTs	3.32 (1.5V)	–	1.67 (2.5 V, 0.1 Hz)
PVDF-HFP/EB [80]	MWCNT/rGO	0.41 (2.0V)	–	0.07 (2.0 V, 0.1 Hz)
TPU/EB [81]	MWCNT/PP	1.43 (2.5V)		0.64 (2.5 V, 0.1 Hz)
Nafion/EB [82]	$pMoS_2$-nSNrGO/ PP	0.25 (1.0V)	10.0	0.88 (1.0 V, 0.1 Hz)
PSS-b-PMB/ $HMImPF_6$ [54]	SWCNT/ PVDF-HFP/EB	–	50.0	0.98 (1.0 V, 0.025 Hz)
PSS-b-PMB/ Imidazole [66]	SWCNT/ PVDF-HFP/EB	0.30 (1.0V)	0.2	0.45 (1.0 V, 0.5 Hz)
PVDF-HFP/EB [83]	SWCNT/PP/EB	–	–	0.57 (2.0 V, 0.05 Hz)
SPBI/EB [84]	Th-SNrGO/PP	1.26 (2.0V)	–	0.36 (1.0 V, 0.1 Hz)
PS-b-PSS/EB [85]	Th-SNrGO/PP	0.27 (2.0V)	5.0	0.56 (1.0 V, 0.1 Hz)
PVDF/EB [68]	Graphdiyne/PVDF/ EB	3.37 (2.5V)	–	0.78 (2.5 V, 0.1 Hz)
PEO-NBR/EB [65]	g-CN/PVDF/EB	0.93 (4.0V)	100.0	0.93 (3.0 V, 0.002 Hz)
Nafion/EB [86]	BS-COF-C900/PP	–	–	0.62 (0.5 V, 0.1 Hz)
Nafion/EB [87]	COFs/PP	2.31 (2.0V)	2.3	0.65 (1.0 V, 0.1 Hz)
G-CNT-Ni/ Nafion/EB [88]	3D G-CNT-Ni/PP	4.53 (2.0V)	–	0.52 (1.0 V, 0.1 Hz)
Nafion/EB [89]	3D GCN-NG/PP	2.45 (2.0V)	–	0.52 (0.5 V, 0.1 Hz)
Nafion/EB [67]	N-doped carbon/PP	1.00 (2.0V)	–	0.53 (0.5 V, 0.1 Hz)
PVDF-HFP/EB [90]	SWCNT/PANI/ PVDF-HFP/EB	–	–	0.72 (2.0 V, 0.1 Hz)
Nafion/EB [91]	Graphene mesh/ NrGO/Nafion	0.30 (1.0V)	7.5	–

(continued)

Table 1.1 (continued)

Electrolyte materials	Electrode materials	Blocking force, mN	Rise time, s	Strain, %
PVDF-HFP/EB [92]	TiC/PVDF-HFP/EB	–	–	0.62 (2.0 V, 0.005 Hz)

significant [9, 47]. Nafion consists of polytetrafluoroethylene backbones for mechanical strength and random perfluoroalkyl ether branches terminated by sulfonate groups for ionic conductivity. While early Nafion membranes incorporated with water produced wet-type actuators [48, 49], later counterparts integrated with ionic liquids generated dry-type actuators [50, 51]. Block copolymers, with their self-assembly ability, enable the manipulation of ionic conducting channels on the nanoscale [52–54].

The original electrodes were platinum, followed by various advanced materials [55]. Among these, graphene, covalent triazine frameworks (CTFs), and MXenes are noteworthy [70, 71, 93]. Graphene, a single layer of carbon atoms, can serve as a substrate for growing capacitive nanomaterials, thereby producing electrode materials with desired properties [94]. CTFs, characterized by their conjugated frameworks and heteroatoms, are highly porous, capacitive, and conductive, making them excellent electrodes [71]. Mxenes are two-dimensional conductors composed of several atomic layers of transition metal carbides, nitrides, and polar surface functional groups, known for their high charge storage capacity. They exhibit enhanced response speed, bending deformation, and blocking forces. However, their practical applications are currently limited by their susceptibility to oxidation and hydrolysis [93].

The mechanism of actuators containing ionic liquids is elaborated in Fig. 1.1a. Applying an electrical field causes ions to rearrange, leading to mass redistribution and consequent bending deformation [2, 85]. Therefore, actuation performance relies on both the velocity and quantity of moving ions (ionic current), influenced by friction and electrical forces. While the former arises from the insufficient connection of the ionic conducting phase within electrolytes, the latter depends on the capacitance and conductivity of electrodes. Thus, material engineering is crucial for developing large, fast, powerful, and durable actuators.

1.1.2 Brief History of Materials Used in Ionic Artificial Muscles

The exploration of rubber's electrical response by Wilhelm Röntgen in 1880 [45] and polyacrylic acid (PAA)'s response to pH by W. Kuhn's group in 1950 [41] were pioneering demonstrations of polymers reacting to environmental stimuli. The contractions of poly(vinyl alcohol-*co*-acrylic acid) (PVA-PAA) in a salt solution through water electrolysis by Hamlen's group in 1965 [40] and polyacrylamide (PAM) under electrical fields by Tanaka's group in 1982 [39] demonstrated the ability of polymers to deform in response to electrical stimuli. The movement of poly(2-acrylamido-2-methylpropane sulfonic acid) (PAMPS) by Osada's group in 1992 [38] and the deformation of biopolymer granules by Fernandez's group in 1993 [42] in response to electricity suggested that polymers could be used as electroactive actuators. Over the span of a century, these discoveries laid the foundations for the development of ionic actuators.

In 1994, Hirai et al. developed the first ionic actuator using PVA gel and two electrodes [46]. This breakthrough spurred significant interest in the field, leading to extensive research on various materials as listed in Fig. 1.1b and Table 1.1. The electrolytes have expanded from PVA [46] to Nafion [55], poly(vinylidene fluoride) (PVDF) [56], poly(VDF-*co*-hexafluoropropylene) (PVDF-HFP) [58], sulfonated poly(styrene-*b*-ethylene-*co*-butylene-*b*-styrene) (SSEBS) [52], poly(butyl styrene-*b*-methylbutylene-*b*-styrene sulfonate-*b*-methylbutylene-*b*-butyl styrene) (SSPB) [53], poly(styrenesulphonate-*b*-methylbutylene) (PSS-b-PMB) [54, 66, 95], natural polymers [60, 96, 97], poly(styrene-*alt*-maleimide) (PSMI)-PVDF blend [59], polyimide [98], Nafion/reduced graphene oxide (rGO) composite [99], poly(ethylene oxide) and nitrile butadiene rubber (PEO/NBR) [65], polybenzimidazole [84], single ion conductors [66], covalent organic frameworks (COFs) [72], and liquid crystal-polymer composites (LCPC) [74]. Similarly, the electrodes have evolved from platinum (Pt) [55] to gold (Au) [57], silver (Ag) [53], polypyrrole (PPy) [56], poly(3,4-ethylenedioxythiophene)-poly(styrenesulfonate) (PP) [63], single-walled carbon nanotubes (SWNTs) [58, 100], multi-walled carbon nanotubes MWNTs/rGO [62], rGO paper [64], graphitic carbon nitride (g-CN) [65, 89], COFs [67, 86, 87, 101], graphdiyne [68], black phosphorous (BP/CNTs) [69], MXenes [70], CTFs [71], metal organic frameworks (MOFs) [73, 75], and biomass porous carbon (BPC) [76]. Among these materials, Nafion, block copolymers, graphene, and CTFs have proven particularly significant for the development of high-performance ionic actuators.

Ionic actuators offer several advantages, including biomimetic compliance similar to natural muscles, low input voltages (less than 5 V), low power consumption, high safety, high blocking force [102], synchronous control [82, 85], and self-sensing capabilities [47, 79, 103]. These features have enabled various demonstrations, particularly in biomimetic robots. Notable examples include jellyfish [77], inchworm [78], soft fingers [71, 82, 87, 101], self-sensing systems [79, 104], butterflies [70], and hummingbirds [73], as illustrated in Fig. 1.1c.

1.1.3 Required Properties of Ionic Artificial Muscles, Electrolyte and Electrode Materials

Soft robotics require ionic actuators to have large bending displacement, short rise time, high blocking force, and long durability, whose meanings and measurements are depicted in Fig. 1.2a. Large bending displacement (*d*) at low input voltages promotes applications and reduces power consumption. Phase delay ($\Delta\Phi$), which is calculated from bending signal, could benefit devices with complex motions because it helps to accurately predict movements. Although $\Delta\Phi$ also indicates response speed, rise time is more commonly used. Short rise time favorably aids high-frequency movements. Blocking force (*f*) reflects actuators' power and helps in selecting appropriate applications. High blocking force could promote many applications. Finally, durability (*c*) reflects actuator lifespan. Obviously, long durability is important to practical applications.

The mechanism illustrated in the inset of Fig. 1.1a suggests that bending displacement and blocking force rely on the number of moving ions; rise time depends on ion velocity. These ion-related properties are associated with friction and electrical forces, which are ultimately determined by material properties. Furthermore, durability depends on material stability and bonding strength between electrodes and electrolytes. Therefore, materials are key to actuation performance.

The friction force is related to ionic conductivity of electrolytes, which requires a conducting component. In addition, robustness needs a non-conducting component

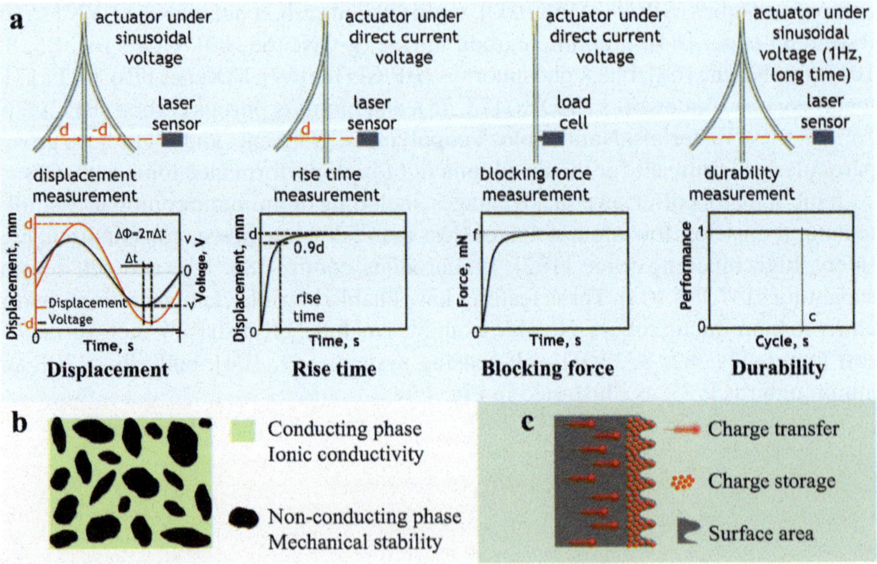

Fig. 1.2 Basic properties of: **a**, actuator; **b**, electrolyte; **c**, electrode

[2, 85]. Furthermore, the ratio and the distribution of these two components in electrolytes directly affect these two properties. For example, a continuous conducting phase preferably generates lower friction force, but unfavorably weakens mechanical strength. Therefore, balancing these two properties in electrolytes is challenging (Fig. 1.2b).

The electrical force is directly linked to electron conductivity and storage capacity of electrodes. Unfortunately, conductivity and capacitance are usually in opposition [82]. Consequently, making good electrodes is difficult. In addition, this force also depends on the ion interaction and accommodation of electrodes, which requires materials with large surface area and high porosity, respectively. Therefore, porous electrodes are preferred (Fig. 1.2c).

1.2 Literature Survey on Electrolyte Materials

As previously analyzed, research on electrolytes should strike a compromise between ionic conductivity and mechanical strength. Consequently, many existing materials have been exploited and new polymers have been synthesized. They can be categorized into six groups: perfluorinated polymers [9, 47], polymer blends [59, 105], polymer composites [99], natural polymers [60, 63], single-ion polymer electrolytes [66], and block isomers [2, 85, 106]. Typical examples for each group are presented in Fig. 1.3.

Perfluorinated polymers' valuable properties are chemical inertness, corrosion resistance, mechanical toughness, thermal stability, and ion conductivity. Some are Nafion [9, 47], Aquivion [107], PVDF [56], PVDF-HFP [58, 100] (Fig. 1.3a). Nafion and Aquivion have similar polar perfluoroalkyl ether branches terminated by strong sulfonic acid groups with pKa of -6 for ionic conductivity and the same polytetrafluoroethylene backbones for robustness. However, Nafion has longer branches and less sulfonic acid groups than Aquivion, which affected actuation performance [107, 108]. PVDF and PVDF-HFP are linear semicrystalline polymers. While amorphous domains mix with ionic liquids for conducting ions, crystalline domains physically bind polymer molecules together for robustness. Trifluoromethyl groups make PVDF-HFP less crystalline, hence more preferable than PVDF. Among those polymers, Nafion shows the best actuation performance as evident in Table 1.1.

Although their high stability hinders these perfluorinated polymers from chemical modification, they can be blended with several polymers. For example, Fig. 1.3b illustrates a bend of poly(styrene sulfonic acid) (PSSA), PVDF, and poly(vinyl pyrrolidone) (PVP) [105]. In this blend, PSSA with sulfonate groups conducted ions, PVDF with high crystallinity provided robustness, and PVP compatibilizer enhanced bonding between these two previous components. However, this blend had lower mechanical properties and less proton conductivity than Nafion, hence modestly enhancing actuation performance. Other blends such as PSMI/PVDF [59], PEO-NBR [65] also showed small actuation improvement.

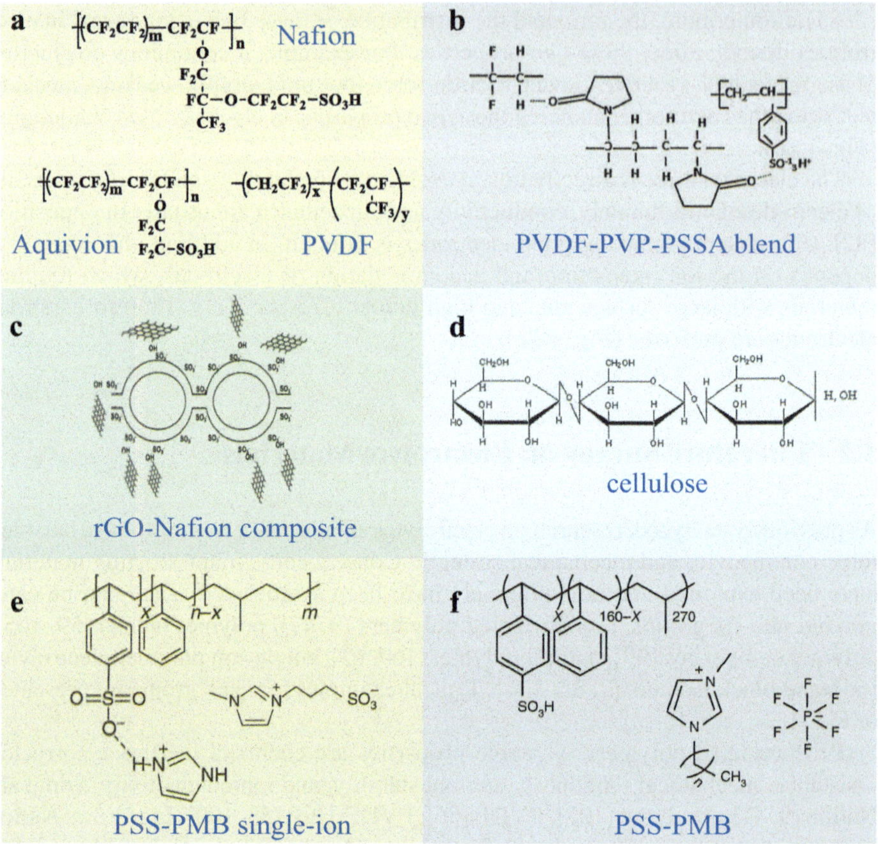

Fig. 1.3 Electrolytes used in ionic actuators: **a** perfluorinated polymers; **b** polymer blends [105]; **c** polymer composites [99]; **d** natural polymers [60] from Refs. [60, 99, 105]. Reprinted with permission from Elsevier; **e** single-ion polymer electrolytes [66]; **f**, block copolymers [54] from Refs. [54, 66]. Reprinted with permission from Springer Nature

A typical polymer composite is presented in Fig. 1.3c [99]. Adding rGO to Nafion enhanced conductivity and robustness, hence somewhat increasing actuation performance. Other composites like Nafion/rGO-CNT-Ni [88], SSPB/GO [109], SSPB/silicate [110, 111], chitosan/GO [96], cellulose/graphene [97, 112] also exhibited small actuation improvement. Therefore, despite the expectation of synergistic effect on increasing both conductivity and robustness, these two properties were not properly and fully addressed and none of these polymer blends and composites can replace Nafion.

Some natural polymers are bacterial cellulose [60, 63, 97], cellulose acetate [112], chitosan [96, 113, 114]. An example is showed in Fig. 1.3d [60]. These polymers are clearly environmental-friendlier than their synthetic counterparts. However, they displayed much poorer performance in comparison with Nafion.

Single-ion polymers can intensify actuation bending because they induce higher mass redistribution than their dual-ion counterparts. Nafion was the first single-ion electrolyte, which contains mobile counter cations and bound sulfonate anions [47, 48, 115]. Many single-ion isomers were investigated [44]. Those single-ion isomers contained water, hence produced wet actuators, which caused two disadvantages. First, wet actuator operating in water prevented applications in open air. Second, the electrolysis of water limited applying voltages less than 1.5 V. Using ionic liquids positively transformed actuators from wet to dry types, but also negatively introduced free cations and anions [50, 51]. Replacing ionic liquids by zwitterions generated a dry single-ion electrolyte, shortening rise time to less than 0.2 s [66] (Fig. 1.3e). However, this electrolyte category suffers from mobile-ion deficiency, hence displays limited deformation [66, 72].

Block isomers could provide a good solution breaking the inverse relationship between ionic conductivity and mechanical strength. While random isomers, like Nafion, irregularly combine ionic and covalent monomer units, causing great difficulty in controlling their morphology, block isomers can separate ionic and covalent units into blocks within a single macromolecule, which can self-assemble into nano-morphologies. Selecting suitable blocks and morphologies could mutually enhance ionic conductivity and mechanical robustness. Commercial materials like SSEBS [52] and SSPB [53] with limited morphological control generated poor performance [2, 85]. Hence, poly(styrene-*b*-methyl methacrylate-*b*-styrene) [116], poly(styrene-*b*-acrylate-*b*-styrene) [117], poly(styrene-*b*-imidazolium bistriflimide-*b*-styrene) (PS-PIB-PS) [118], and poly(styrenesulfonate-*b*-methylbutylene) (PSS-PMB) [54, 66, 95] were synthesized, which partly separated conducting from nonconducting blocks and somewhat controlled morphology. For example, PS-PIB-PS provided no morphology data; conducting block of PSS-PMB (Fig. 1.3f) contained at least 25% nonconducting units [54, 118]. Another challenge is to control self-assembly for continuous conducting phase in hundred-micrometer-thick films because all available techniques for governing self-assembly of block copolymers limit to nano thin films.

This systematic literature survey, categorization, and analyses of electrolyte materials strongly highlight Nafion and block isomers prominent. However, current studies of Nafion, which directly mix with ionic liquids [70, 71, 82, 87] and are simply modified [88, 99, 119], do not pay adequate attentions on its phase separation, hence hindering it from showing its best properties. Recently, theoretical [120] and experimental [121] data have proven different configurations of Nafion according to solvent media. Hence, its morphology could be manipulated for high ionic conductivity and adequate mechanical strength. Because the aforementioned block isomers failed to completely control conducting phase in micrometer-thick films, new materials and novel techniques for governing phase development are required.

1.3 Literature Survey on Electrode Materials

As aforementioned, studies on electrodes should balance between conductance and storage of electrons. Consequently, many existing electrodes have been employed and new materials have been developed. They can be classified into three groups: single materials, composites with and without PP. Representative examples for each group are summarized in Fig. 1.4.

Single materials includes noble metals (Pt [55], Au [57] and Ag [53]), conducting polymers (PPy [56] and PP [63]), CNTs [58, 100], rGO [64], g-CN [65, 89], graphdiyne [68], etc. (Fig. 1.4a). Despite their easy use, they have many drawbacks. Noble metals suffered from cracking [122], inducing poor durability. This was solved by conducting polymers, CNTs, and rGO with high electrical conductivity. However, their limited capacitance produced small deformation. In contrast, capacitive materials like g-CN and graphdiyen have low electrical conductivity, hence generated slow response. Therefore, composites of conducting and capacitive components have been manipulated.

Composites without PP includes rGO and silver nanoparticle (AgNP) [123], rGO/CNTs [80], graphene mesh and nitrogen doped rGO (NrGO) [91], BP/CNTs [69], sulfur-doped nickel cobalt selenide grown on carbon cloth (S-(NiCo)Se/CC) [124], etc. (Fig. 1.4b), of which, rGO, CNTs, graphene mesh, and carbon cloth provided conductivity; AgNP [123], CNTs [80], NrGO [91], BP [69], and S-(NiCo)Se [124] nanofillers improved capacitance. These composites usually require polymer binders for fabricating electrode films. Although this approach is straightforward, developing technologies for regular dispersions of conducting and capacitive materials in polymer matrixes is challenging.

Composites with PP significantly enhanced actuation performance. PP serves three functions as a binder, a charge conductor via PEDOT, and an ion conductor through PSS. In addition, its well dispersion in water easily mixes with nanofillers, and its versatile cast film facilitates actuator fabrication. Many nanofillers were used. Typical examples Fig. 1.4c are rGO/CNTs [88], graphene/Ag nanowires [125], MXenes [93], and CTFs [71]. The high capacitance and large surface area of nanofillers can be achieved by several techniques, which include preventing restacking of graphene by growing CNTs on its sheets [88], using 2D nanomaterials like MXenes with reversible ion intercalations [93], and developing stable capacitive porous materials like CTFs [71]. These CTFs have good thermal and chemical stability due to their triazine and benzene rings, high capacitance due to their conjugated systems and rich heteroatoms, and enormous porosity due to their framework. Hence, composite electrodes of CTFs with PP showed significant actuation improvement [71]. A small stimulating voltage of 0.5 V made actuators significantly bend up to 17 mm.

This organized literature review, classification, and investigation of electrode materials strongly recommend graphene nanocomposite and CTFs to be imperative. While MXenes have limitations of stability, graphene nanocomposite [127] and CTFs [128] with high stability could be promising electrode candidates, providing great

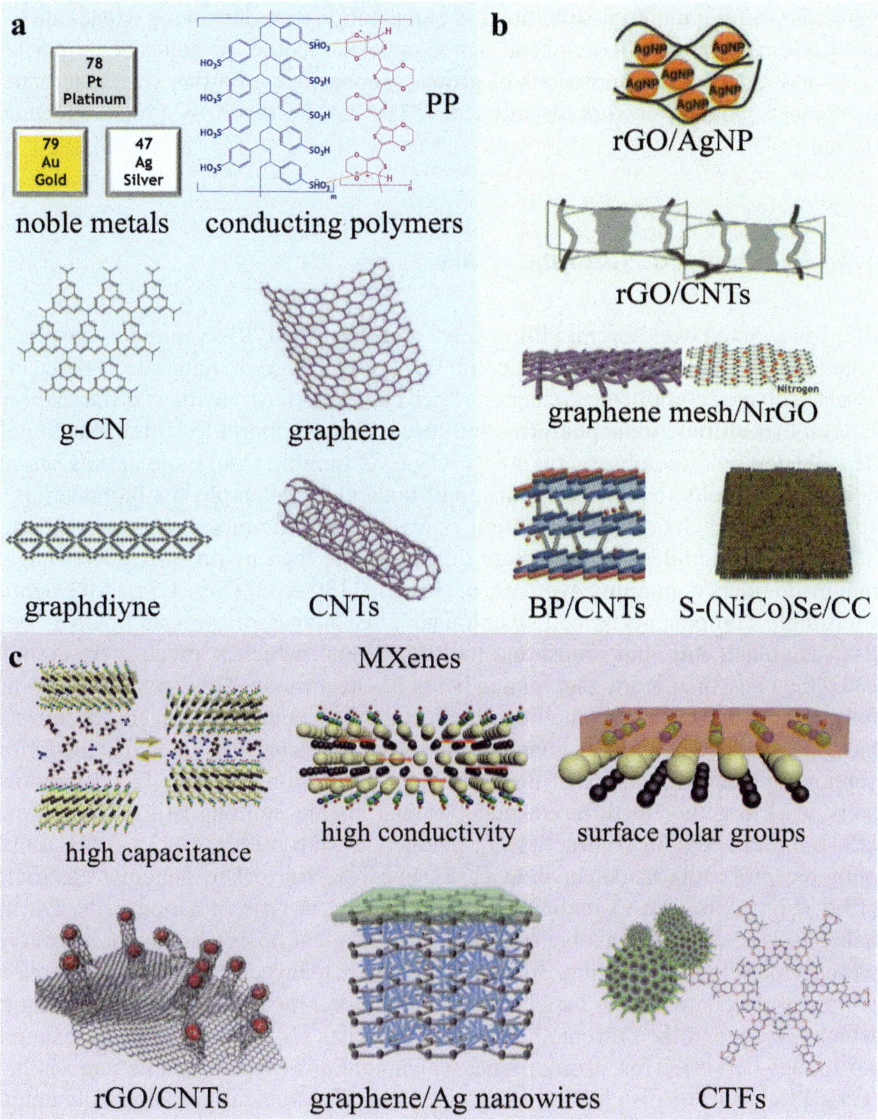

Fig. 1.4 Electrodes used in ionic actuators: **a** single materials, noble metals, conducting polymers (PP), g-CN [65], graphene [126], graphidiyne [68], CNTs [126] from Refs. [68, 126]. Reprinted with permission from Springer Nature; **b** composites without PP, rGO/AgNP [123], rGO/CNTs [80], graphene mesh/NrGO [91], BP/CNTs [69] from Refs. [69, 80, 91, 123]. Reprinted with permission from John Wiley and Sons, S-(NiCa)Se/CC [124] from Ref. [124]. Reprinted with permission from ACS Publications; **c** composites with PP, MXenes, rGO/CNTs [88] from Ref. [88]. Reprinted with permission from John Wiley and Sons, graphene/Ag nanowires [125] from Ref. [152]. Reprinted with permission from MDPI, CTFs [71] from Ref. [71]. Reprinted with permission from Springer Nature

versatility tuning material structure and compositions for improving capacitance of electrodes. Properties of graphene nanocomposites could be adjusted by particle sizes and chemical compositions of grown nanoparticles. Porous size, heteroatom percentage, and framework structure of CTFs can be tuned by proper precursor design.

1.4 Bioinspired Nanomaterials

Life has evolved over several billion years on the Earth [129]. Nature has developed ingenious methods that delicately combine several common materials at molecular levels in diverse complicated patterns, which carefully integrate many different properties into multifunctional platforms and successfully support life. This can definitely be validated by close observations on every level ranging from biomaterials, simple components, cells to tissues, organs, and bodies. An example for biomaterials is spider silk. This delicate combination of crystalline and amorphous regions, made of natural polyamides, is over seven times tougher than nylon fiber, the strongest manmade fibers containing synthetic polyamide [130–132] (Fig. 1.5a). An example for simple components is cell membrane. This bilayer of phospholipids is only five nanometer thin, but robust and multifunctional, which is much more complicated than any membrane that human being has ever made. This membrane can not only protect cells but also facilitate their metablism. Furthermore, this membrane has many selective protein channels that precisely control material penetration, supporting many functions of different cells. Many ion channels [133] harmonically work together in different manners that enable neurons to transmit signals [134], muscle cells to contract [135], platelets to stop bleeding [136, 137], retinal photoreceptor cells to detect light [138], even electric eel to generate electricity [139] (Fig. 1.5b). An example for cells is platele, a type of blood cells. During hemostasis, these cells involve sequential biochemical processes that quickly form strong clots to stop bleeding. In very short time, many electrochemical reactions are precisely controlled to happen right in the blood medium at body temperature, which are beyond the current ability of all chemists [140] (Fig. 1.5c). An example for tissues is bone. This strong tissue, which contains hydroxyapatite nanocrystals ($Ca_{10}(PO_4)_6(OH)_2$) distributing in collagen, has hierarchical porosity. This unique structure provides spaces for blood vessels, cells, inorganic minerals, bone marrow, and so on [141–145], which serve diverse functions. In addition, this porous structure makes bones very light but extremely strong, which supports a frame for whole body and protects soft organs (Fig. 1.5d). Furthermore, bones and skeletal muscles form living actuators responsible for all motion functions of animals. Surprisingly, these soft actuators convert chemical to mechanical energy with efficiency higher than that of typical car engine [1]. An interdisciplinary group containing excellent experts in biology, chemistry, materials, mechanics could find great difficulty in developing a structure like a simple bone.

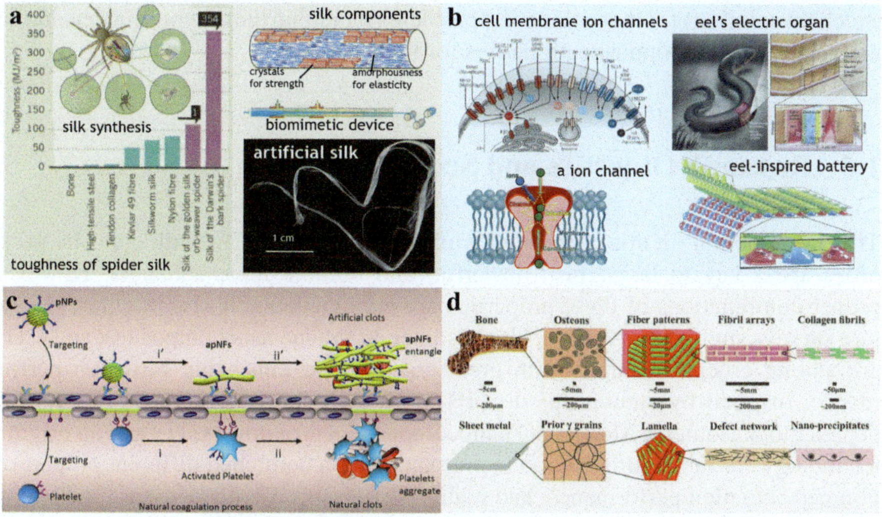

Fig. 1.5 Bioinspired nanomaterials: **a** spider silk [130, 146] and artificial silk [147] from Refs. [130, 146, 147]. Reprinted with permission from Springer Nature; **b** cell membrane ion channels [133, 148] from Ref. [133]. Reprinted with permission from AAAS, from Ref. [148]. Reprinted with permission from ACS Publications, and eel-inspired battery [139] from Ref. [139]. Reprinted with permission from Springer Nature; **c** platelet cell and its mimetic artificial coagulation [140] from Ref. [140]. Reprinted with permission from AAAS; **d** bone structure and its inspired hierarchical metastable steel [149] from Ref. [149]. Reprinted with permission from AAAS

Recent rapid advance in biology has clarified many concepts, structures, compositions, mechanisms, and functions of various biocomponents. This has provided scientists and engineers excellent solutions and tools for many technical and practical problems relating to energy efficiency, multifunctional materials, health, clean water, and food security [129, 150, 151]. Bioinspiration and biomimetic research has expanded into architectural design, materials, chemistry, mechanics, robotics, engineering, and many technologies [152]. Complex biocomposites have encouraged new designs and combinations of current materials to generate better properties [143, 153–155]. Spider silk has motivated scientists to develop extremely robust materials [130, 131, 146] and artificial silk [147] (Fig. 1.5a). Selective channels in cell membrane inspired a pore design in membranes for desalinization [156] and batteries [139] (Fig. 1.5b). Platelet inspired artificial coagulation [140] (Fig. 1.5c). Bone structure prompted hierarchical metastable steel [149] (Fig. 1.5d). The adhesive proteins in mussels and the foot-hair pattern of geckos inspired reversible adhesives [157–159]. Biological optics stimulated advanced photonic systems [160]. Octopus triggered a new hot field, which is soft robotics [161].

Despite its current flourish, the field of bioinspirations and biomimetics has just started in few species like geckos, mussels, octopus, and several materials like spider silk, bones, and several proteins. There are countless species, materials, and structures awaiting to be discovered and mimicked. Investigating this toolbox could provide excellent solutions for many scientific problems including the dilemma in developing nanomaterials with opposite properties for ionic actuators.

1.5 Research Objective and Scope

This Introduction (in this chapter) identifies that nanomaterials with opposite properties are a key to large, fast, powerful, and durable ionic actuators. However, proper combinations of these properties are very challenging. Furthermore, literature surveys suggest that Nafion, block isomer, graphene nanocomposites, and CTFs are among crucial materials with great potential for enhancing actuation performance. Inspired by nature, this dissertation presents three materials (an unprecedented block isomer, a MoS_2-rGO nanocomposite, and a novel CTF) and an original technology for fabricating Nafion membrane, which combined different properties, boomed actuation performance, and realized some soft robotic demonstrations.

Chapter 2 presents an unprecedented cell-membrane-inspired polystyrene-*b*-poly(1-ethyl-3-methylimidazolium-4-styrenesulfonate) (PS-*b*-PSS-EMIm) block isomer with hydrophobic PS cores for mechanical strength, hydrophilic PSS shells for ionic conductivity, and EMIm cation for free ion migration, which respectively resemble to hydrophobic tail, hydrophilic head, and calcium cation of phospholipid in cell membrane [162] (Fig. 1.6a). This special structure enabled to form micro-thick electrolyte membranes with continuous conducting phase for high ionic conductivity and separated non-conducting domains for adequate mechanical strength, hence producing good actuation performance.

Chapter 3 presents an original hemostasis-inspired Nafion membrane with high ionic conductivity and adequate mechanical strength [136, 137] (Fig. 1.6b). A proper solvent system induced Nafion micelles with conducting shells covering non-conducting cores. This micellar structure is similar to that of cell membrane. Casting these micelles formed robust electrolyte membrane, resembling hemostasis. This membrane had continuous conducting phase for high ionic conductivity and well-connected non-conducting domains for good mechanical strength. Therefore, actuators showed displacement improvement by 35 times (0.3 V and 3 Hz) and realized an inchworm-mimetic robot and a kinetic tensegrity system.

Chapter 4 presents a nanocomposite material that fast transferred and stored electrons, inspired by the quick calcium switch between inactive state stored in bones and active state in the blood and cells. MoS_2 was selected to store charges because of its high specific capacitance, which resembles bones storing calcium. Graphene was chosen to conduct charge due to its excellent conductivity, which mimics blood vessels transporting calcium in bones and whole body. Furthermore, the large contact area between these two components resembles dense tiny blood vessels in bones (Fig. 1.6c). Applying this bioinspired MoS_2-rGO nanocomposite as electrode in ionic actuators multiplied actuation bending by 6.5 times in comparison with rGO at 0.5 V and 1 Hz. Therefore, the developed ionic actuators enabled soft fingers to interact with delicate display surfaces of smartphones and tablets.

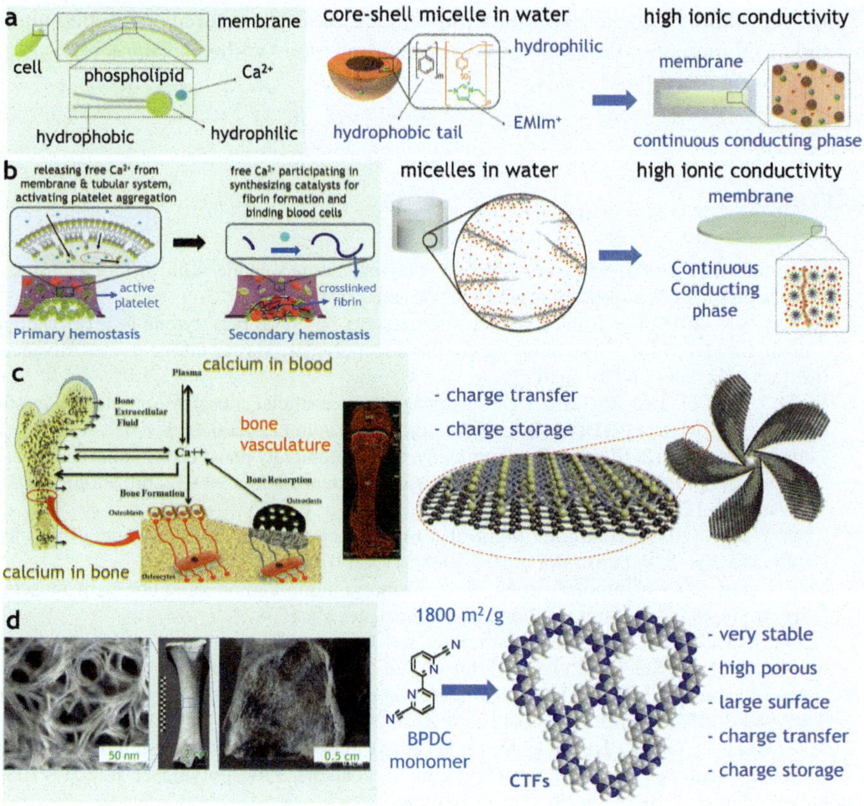

Fig. 1.6 Research objectives: **a** cell-membrane-inspired block copolymer electrolyte [85] from Ref. [85]. Reprinted with permission from John Wiley and Sons; **b** hemostasis-inspired electrolyte membrane [163] from Ref. [163]. Reprinted with permission from Springer Nature; **c** MoS_2-rGO nano composite inspired by calcium stored and transported in the body [82, 164, 165] from Ref. [82]. Reprinted with permission from John Wiley and Sons, from Ref. [164]. Reprinted with permission from the Company of Biologists, from Ref. [165]. Reprinted with permission from Springer Nature; **d** bone-structure-inspired hierarchical porous CTF electrode, bone structures [144, 145, 166] from Ref. [144]. Reprinted with permission from Annual Reviews, from Ref. [145]. Reprinted with permission from Springer Nature, from Ref. [166]. Reprinted with permission from John Wiley and Sons

Chapter 5 introduces a novel bone-structure-inspired CTF derived from 2,2′-Bipyridine-3,3′-dicarbonitrile [144, 145]. This CTF was designed to have hierarchical porosity, high surface area, a large amount of nitrogen heteroatom, and partial carbonization, which altogether contributed to high capacitance (Fig. 1.6d). Furthermore, its hierarchical porosity facilitated ion penetration into it, promoting strong interaction between this CTF and ions during actuator operation. Therefore, adding about 4 wt% to PP generated good electrodes, which multiplied bending deformation over 2.5 times at 0.5 V and 1.0 Hz, and enabled dragonfly-mimetic demonstration.

The final Chap. 6 summarizes main importance about these three developed materials and the Nafion membrane technology including their synthesis, characterization, and contribution to related fields.

References

1. Baughman RH (2005) Materials science. Playing nature's game with artificial muscles. Science 308:63–65. https://doi.org/10.1126/science.1099010
2. Nguyen VH (2018) Electromechanical soft actuators based on polystyrene-b-poly (1-alkyl-3-methylimidazolium p-styrenesulfonate) block copolymer. Master thesis, Korea Advanced Institute of Science and Technology
3. Wang J, Gao D, Lee PS (2020) Recent progress in artificial muscles for interactive soft robotics. Adv Mater e2003088. https://doi.org/10.1002/adma.202003088
4. Lima MD et al (2012) Electrically, chemically, and photonically powered torsional and tensile actuation of hybrid carbon nanotube yarn muscles. Science 338:928–932. https://doi.org/10.1126/science.1226762
5. Chen Y et al (2019) Controlled flight of a microrobot powered by soft artificial muscles. Nature 575:324–329. https://doi.org/10.1038/s41586-019-1737-7
6. Chu H et al (2021) Unipolar stroke, electroosmotic pump carbon nanotube yarn muscles. Science 371:494–498. https://doi.org/10.1126/science.abc4538
7. Acome E et al (2018) Hydraulically amplified self-healing electrostatic actuators with muscle-like performance. Science 359:61–65. https://doi.org/10.1126/science.aao6139
8. Jager EW, Smela E, Inganas O (2000) Microfabricating conjugated polymer actuators. Science 290:1540–1545. https://doi.org/10.1126/science.290.5496.1540
9. Jo C, Pugal D, Oh IK, Kim KJ, Asaka K (2013) Recent advances in ionic polymer-metal composite actuators and their modeling and applications. Prog Polym Sci 38:1037–1066. https://doi.org/10.1016/j.progpolymsci.2013.04.003
10. Gao D et al (2023) A supramolecular gel-elastomer system for soft iontronic adhesives. Nat Commun 14:1990. https://doi.org/10.1038/s41467-023-37535-4
11. Kim Y, Yuk H, Zhao R, Chester SA, Zhao X (2018) Printing ferromagnetic domains for untethered fast-transforming soft materials. Nature 558:274–279. https://doi.org/10.1038/s41586-018-0185-0
12. Hu W, Lum GZ, Mastrangeli M, Sitti M (2018) Small-scale soft-bodied robot with multimodal locomotion. Nature 554:81–85. https://doi.org/10.1038/nature25443
13. Zhang YX et al (2023) Coaxially printed magnetic mechanical electrical hybrid structures with actuation and sensing functionalities. Nat Commun 14:4428. https://doi.org/10.1038/s41467-023-40109-z
14. Iwaso K, Takashima Y, Harada A (2016) Fast response dry-type artificial molecular muscles with [c2]daisy chains. Nat Chem 8:625–632. https://doi.org/10.1038/nchem.2513
15. Chen J et al (2018) Artificial muscle-like function from hierarchical supramolecular assembly of photoresponsive molecular motors. Nat Chem 10:132–138. https://doi.org/10.1038/nchem.2887
16. Lendlein A, Jiang H, Junger O, Langer R (2005) Light-induced shape-memory polymers. Nature 434:879–882. https://doi.org/10.1038/nature03496
17. Beebe DJ et al (2000) Functional hydrogel structures for autonomous flow control inside microfluidic channels. Nature 404:588–590. https://doi.org/10.1038/35007047
18. Dong L, Agarwal AK, Beebe DJ, Jiang H (2006) Adaptive liquid microlenses activated by stimuli-responsive hydrogels. Nature 442:551–554. https://doi.org/10.1038/nature05024
19. Walters DH, Kuhn W, Kuhn HJ (1961) Action potential with an artificial pH-muscle. Nature 189:381–383. https://doi.org/10.1038/189381a0

20. Kim SJ et al (2006) Surprising shrinkage of expanding gels under an external load. Nat Mater 5:48–51. https://doi.org/10.1038/nmat1553
21. Steinberg IZ, Oplatka A, Katchalsky A (1966) Mechanochemical engines. Nature 210:568–571. https://doi.org/10.1038/210568a0
22. Lendlein A, Gould OEC (2019) Reprogrammable recovery and actuation behaviour of shape-memory polymers. Nat Rev Mater 4:116–133. https://doi.org/10.1038/s41578-018-0078-8
23. Haines CS et al (2014) Artificial muscles from fishing line and sewing thread. Science 343:868–872. https://doi.org/10.1126/science.1246906
24. Kanik M et al (2019) Strain-programmable fiber-based artificial muscle. Science 365:145–150. https://doi.org/10.1126/science.aaw2502
25. Downs FG et al (2020) Multi-responsive hydrogel structures from patterned droplet networks. Nat Chem 12:363–371. https://doi.org/10.1038/s41557-020-0444-1
26. Ma M, Guo L, Anderson DG, Langer R (2013) Bio-inspired polymer composite actuator and generator driven by water gradients. Science 339:186–189. https://doi.org/10.1126/science.1230262
27. Shin B et al (2018) Hygrobot: a self-locomotive ratcheted actuator powered by environmental humidity. Sci Robot 3:eaar2629. https://doi.org/10.1126/scirobotics.aar2629
28. Hu Z, Zhang X, Li Y (1995) Synthesis and application of modulated polymer gels. Science 269:525–527. https://doi.org/10.1126/science.269.5223.525
29. Katzschmann RK, DelPreto J, MacCurdy R, Rus D (2018) Exploration of underwater life with an acoustically controlled soft robotic fish. Sci Robot 3. https://doi.org/10.1126/scirobotics.aar3449
30. Yuk H et al (2017) Hydraulic hydrogel actuators and robots optically and sonically camouflaged in water. Nat Commun 8:14230. https://doi.org/10.1038/ncomms14230
31. Siefert E, Reyssat E, Bico J, Roman B (2019) Bio-inspired pneumatic shape-morphing elastomers. Nat Mater 18:24–28. https://doi.org/10.1038/s41563-018-0219-x
32. Bartlett NW et al (2015) Soft Robotics. A 3D-printed, functionally graded soft robot powered by combustion. Science 349:161–165. https://doi.org/10.1126/science.aab0129
33. Wehner M et al (2016) An integrated design and fabrication strategy for entirely soft, autonomous robots. Nature 536:451. https://www.nature.com/articles/nature19100#supplementary-information, https://doi.org/10.1038/nature19100
34. Xu C, Stiubianu GT, Gorodetsky AA (2018) Adaptive infrared-reflecting systems inspired by cephalopods. Science 359:1495–1500. https://doi.org/10.1126/science.aar5191
35. Zhang QM et al (2002) An all-organic composite actuator material with a high dielectric constant. Nature 419:284–287. https://doi.org/10.1038/nature01021
36. Kwon IC, Bae YH, Kim SW (1991) Electrically erodible polymer gel for controlled release of drugs. Nature 354:291–293. https://doi.org/10.1038/354291a0
37. Bar-Cohen Y (2004) SPIE. The International Society for Optical Engineering
38. Osada Y, Okuzaki H, Hori H (1992) A polymer gel with electrically driven motility. Nature 355:242–244. https://doi.org/10.1038/355242a0
39. Tanaka T, Nishio I, Sun ST, Ueno-Nishio S (1982) Collapse of gels in an electric field. Science 218:467–469. https://doi.org/10.1126/science.218.4571.467
40. Hamlen RP, Kent CE, Shafer SN (1965) Electrolytically activated contractile polymer. Nature 206:1149–1150. https://doi.org/10.1038/2061149b0
41. Kuhn W, Hargitay B, Katchalsky A, Eisenberg H (1950) Reversible dilation and contraction by changing the state of ionization of high-polymer acid networks. Nature 165:514–516. https://doi.org/10.1038/165514a0
42. Nanavati C, Fernandez JM (1993) The secretory granule matrix: a fast-acting smart polymer. Science 259:963–965. https://doi.org/10.1126/science.8438154
43. Carrico JD, Tyler T, Leang KK (2018) A comprehensive review of select smart polymeric and gel actuators for soft mechatronics and robotics applications: fundamentals, freeform fabrication, and motion control. Int J Smart Nano Mater 8:144–213. https://doi.org/10.1080/19475411.2018.1438534

44. Bahramzadeh Y, Shahinpoor M (2014) A review of ionic polymeric soft actuators and sensors. Soft Robot 1:38–52. https://doi.org/10.1089/soro.2013.0006
45. Röntgen WC (1880) Ueber die durch Electricität bewirkten Form- und Volumenänderungen von dielectrischen Körpern. Annalen der Physik und Chemie 247:771–786. https://doi.org/10.1002/andp.18802471304
46. Hirai T, Nemoto H, Hirai M, Hayashi S (1994) Electrostriction of highly swollen polymer gel-possible application for cell actuator. J Appl Polym Sci 53:79–84. https://doi.org/10.1002/app.1994.070530109
47. Shahinpoor M, Bar-Cohen Y, Simpson JO, Smith J (1998) Ionic polymer-metal composites (IPMCs) as biomimetic sensors, actuators and artificial muscles-a review. Smart Mater Struct 7:R15–R30. https://doi.org/10.1088/0964-1726/7/6/001
48. Jeon J-H, Yeom S-W, Oh I-K (2008) Fabrication and actuation of ionic polymer metal composites patterned by combining electroplating with electroless plating. Compos A Appl Sci Manuf 39:588–596. https://doi.org/10.1016/j.compositesa.2007.07.013
49. Madden JDW et al (2004) Artificial muscle technology: physical principles and naval prospects. IEEE J Oceanic Eng 29:706–728. https://doi.org/10.1109/Joe.2004.833135
50. Lee JW, Yoo YT, Lee JY (2014) Ionic polymer-metal composite actuators based on triple-layered polyelectrolytes composed of individually functionalized layers. ACS Appl Mater Interfaces 6:1266–1271. https://doi.org/10.1021/am405090d
51. Liu Y et al (2013) Direct observation of ion distributions near electrodes in ionic polymer actuators containing ionic liquids. Sci Rep 3:973. https://doi.org/10.1038/srep00973
52. Wang XL, Oh IK, Lu J, Ju JH, Lee S (2007) Biomimetic electro-active polymer based on sulfonated poly (styrene-b-ethylene-co-butylene-b-styrene). Mater Lett 61:5117–5120. https://doi.org/10.1016/j.matlet.2007.04.004
53. Vargantwar PH, Shankar R, Krishnan AS, Ghosh TK, Spontak RJ (2011) Exceptional versatility of solvated block copolymer/ionomer networks as electroactive polymers. Soft Matter 7:1651. https://doi.org/10.1039/c0sm01210f
54. Kim O, Shin TJ, Park MJ (2013) Fast low-voltage electroactive actuators using nanostructured polymer electrolytes. Nat Commun 4:2208. https://doi.org/10.1038/ncomms3208
55. Asaka K, Oguro K, Nishimura Y, Mizuhata M, Takenaka H (1995) Bending of polyelectrolyte membrane–platinum composites by electric stimuli I. Response characteristics to various waveforms. Polym J 27:436–440. https://doi.org/10.1295/polymj.27.436
56. Zhou DZ et al (2003) Solid state actuators based on polypyrrole and polymer-in-ionic liquid electrolytes. Electrochim Acta 48:2355–2359. https://doi.org/10.1016/S0013-4686(03)00225-1
57. Oguro K, Fujiwara N, Asaka K, Onishi K, Sewa S (1999) Polymer electrolyte actuator with gold electrodes. Proc Spie 3669:64–71. https://doi.org/10.1117/12.349698
58. Fukushima T, Asaka K, Kosaka A, Aida T (2005) Fully plastic actuator through layer-by-layer casting with ionic-liquid-based bucky gel. Angew Chem Int Ed Engl 44:2410–2413. https://doi.org/10.1002/anie.200462318
59. Lu J, Kim SG, Lee S, Oh IK (2008) A biomimetic actuator based on an ionic networking membrane of poly(styrene-alt-maleimide)-incorporated poly(vinylidene fluoride). Adv Funct Mater 18:1290–1298. https://doi.org/10.1002/adfm.200701133
60. Jeon JH, Oh IK, Kee CD, Kim SJ (2010) Bacterial cellulose actuator with electrically driven bending deformation in hydrated condition. Sensor Actuat B-Chem 146:307–313. https://doi.org/10.1016/j.snb.2010.02.046
61. Jung JH, Jeon JH, Sridhar V, Oh IK (2011) Electro-active graphene-Nafion actuators. Carbon 49:1279–1289. https://doi.org/10.1016/j.carbon.2010.11.047
62. Yang W, Choi H, Choi S, Jeon M, Lee SY (2012) Carbon nanotube-graphene composite for ionic polymer actuators. Smart Mater Struct 21:055012. https://doi.org/10.1088/0964-1726/21/5/055012
63. Kim SS, Jeon JH, Kee CD, Oh IK (2013) Electro-active hybrid actuators based on freeze-dried bacterial cellulose and PEDOT:PSS. Smart Mater Struct 22:085026. https://doi.org/10.1088/0964-1726/22/8/085026

64. Kim J, Jeon JH, Kim HJ, Lim H, Oh IK (2014) Durable and water-floatable ionic polymer actuator with hydrophobic and asymmetrically laser-scribed reduced graphene oxide paper electrodes. ACS Nano 8:2986–2997. https://doi.org/10.1021/nn500283q
65. Wu G et al (2015) Graphitic carbon nitride nanosheet electrode-based high-performance ionic actuator. Nat Commun 6:7258. https://doi.org/10.1038/ncomms8258
66. Kim O, Kim H, Choi UH, Park MJ (2016) One-volt-driven superfast polymer actuators based on single-ion conductors. Nat Commun 7:13576. https://doi.org/10.1038/ncomms13576
67. Roy S, Kim J, Kotal M, Kim KJ, Oh IK (2017) Electroionic antagonistic muscles based on nitrogen-doped carbons derived from poly(triazine-triptycene). Adv Sci (Weinh) 4:1700410. https://doi.org/10.1002/advs.201700410
68. Lu C et al (2018) High-performance graphdiyne-based electrochemical actuators. Nat Commun 9:752. https://doi.org/10.1038/s41467-018-03095-1
69. Wu G et al (2019) High-performance hierarchical black-phosphorous-based soft electrochemical actuators in bioinspired applications. Adv Mater 31:e1806492. https://doi.org/10.1002/adma.201806492
70. Umrao S et al (2019) MXene artificial muscles based on ionically cross-linked $Ti_3C_2T_x$ electrode for kinetic soft robotics. Sci Robot 4:eaaw7797. https://doi.org/10.1126/scirobotics.aaw7797
71. Mahato M et al (2020) CTF-based soft touch actuator for playing electronic piano. Nat Commun 11:5358. https://doi.org/10.1038/s41467-020-19180-3
72. Yu F et al (2022) Ionic covalent organic framework based electrolyte for fast-response ultra-low voltage electrochemical actuators. Nat Commun 13:390. https://doi.org/10.1038/s41467-022-28023-2
73. Mahato M et al (2022) A dual-responsive magnetoactive and electro-ionic soft actuator derived from a nickel-based metal-organic framework. Adv Mater 34:e2203613. https://doi.org/10.1002/adma.202203613
74. Liu C, Cao S, Yoshio M (2023) Ion-conducting non-flammable liquid crystal–polymer composites for high-frequency soft actuators. Adv Funct Mater 33. https://doi.org/10.1002/adfm.202300538
75. Taseer AK et al (2024) Cobalt MOF-based porous carbonaceous spheres for multimodal soft actuator exhibiting intricate biomimetic motions. Adv Mater 36:e2312340. https://doi.org/10.1002/adma.202312340
76. Zhang H et al (2024) Soft actuator with biomass porous electrode: a strategy for lowering voltage and enhancing durability. Nano Lett. https://doi.org/10.1021/acs.nanolett.4c01129
77. Yeom SW, Oh IK (2009) A biomimetic jellyfish robot based on ionic polymer metal composite actuators. Smart Mater Struct 18:085002. https://doi.org/10.1088/0964-1726/18/8/085002
78. Must I et al (2015) Ionic and capacitive artificial muscle for biomimetic soft robotics. Adv Eng Mater 17:84–94. https://doi.org/10.1002/adem.201400246
79. Tabassian R et al (2019) Self-sensing actuators: graphene mesh for self-sensing ionic soft actuator inspired from mechanoreceptors in human body. Adv Sci 6(23). https://doi.org/10.1002/advs.201970137
80. Lu L et al (2012) Highly stable air working bimorph actuator based on a graphene nanosheet/carbon nanotube hybrid electrode. Adv Mater 24:4317–4321. https://doi.org/10.1002/adma.201201320
81. Wang DX et al (2017) High energy conversion efficiency conducting polymer actuators based on PEDOT:PSS/MWCNTs composite electrode. Rsc Adv 7:31264–31271. https://doi.org/10.1039/c7ra05469f
82. Manzoor MT et al (2019) Mutually exclusive p-type and n-type hybrid electrode of MoS_2 and graphene for artificial soft touch fingers. Adv Funct Mater. https://doi.org/10.1002/adfm.201905454
83. Terasawa N, Asaka K (2016) High-performance PEDOT:PSS/single-walled carbon nanotube/ionic liquid actuators combining electrostatic double-layer and faradaic capacitors. Langmuir 32:7210–7218. https://doi.org/10.1021/acs.langmuir.6b01148

84. Kotal M, Kim J, Kim KJ, Oh IK (2016) Sulfur and nitrogen co-doped graphene electrodes for high-performance ionic artificial muscles. Adv Mater 28:1610–1615. https://doi.org/10.1002/adma.201505243
85. Nguyen VH et al (2019) Electroactive artificial muscles based on functionally antagonistic core-shell polymer electrolyte derived from PS-b-PSS block copolymer. Adv Sci (Weinh) 6:1801196. https://doi.org/10.1002/advs.201801196
86. Roy S et al (2019) Collectively exhaustive electrodes based on covalent organic framework and antagonistic co-doping for electroactive ionic artificial muscles. Adv Funct Mater. https://doi.org/10.1002/adfm.201900161
87. Mahato M et al (2020) Sulfur- and nitrogen-rich porous π-conjugated COFs as stable electrode materials for electro-ionic soft actuators. Adv Funct Mater 30. https://doi.org/10.1002/adfm.202003863
88. Kim J et al (2017) Soft but powerful artificial muscles based on 3D graphene-CNT-Ni heteronanostructures. Small 13. https://doi.org/10.1002/smll.201701314
89. Kotal M et al (2018) Highly bendable ionic soft actuator based on nitrogen-enriched 3D hetero-nanostructure electrode. Adv Funct Mater 28:1802464. https://doi.org/10.1002/adfm.201802464
90. Sugino T, Kiyohara K, Takeuchi I, Mukai K, Asaka K (2009) Actuator properties of the complexes composed by carbon nanotube and ionic liquid: the effects of additives. Sens Actuators B Chem 141:179–186. https://doi.org/10.1016/j.snb.2009.06.002
91. Tabassian R, Kim J, Nguyen VH, Kotal M, Oh I-K (2018) Functionally antagonistic hybrid electrode with hollow tubular graphene mesh and nitrogen-doped crumpled graphene for high-performance ionic soft actuators. Adv Funct Mater 28:1705714. https://doi.org/10.1002/adfm.201705714
92. Torop J et al (2011) Flexible supercapacitor-like actuator with carbide-derived carbon electrodes. Carbon 49:3113–3119. https://doi.org/10.1016/j.carbon.2011.03.034
93. Nguyen VH et al (2020) Stimuli-responsive MXene-based actuators. Adv Funct Mater 30:1909504. https://doi.org/10.1002/adfm.201909504
94. Anasori B, Beidaghi M, Gogotsi Y (2014) Graphene–transition metal oxide hybrid materials. Mater Today 17:253–254. https://doi.org/10.1016/j.mattod.2014.04.043
95. Kim O, Kim SY, Park B, Hwang W, Park MJ (2014) Factors affecting electromechanical properties of ionic polymer actuators based on ionic liquid-containing sulfonated block copolymers. Macromolecules 47:4357–4368. https://doi.org/10.1021/ma500869h
96. Jeon JH, Cheedarala RK, Kee CD, Oh IK (2013) Dry-type artificial muscles based on pendent sulfonated chitosan and functionalized graphene oxide for greatly enhanced ionic interactions and mechanical stiffness. Adv Funct Mater 23:6007–6018. https://doi.org/10.1002/adfm.201203550
97. Kim SS, Jeon JH, Kim HI, Kee CD, Oh IK (2015) High-fidelity bioelectronic muscular actuator based on graphene-mediated and TEMPO-oxidized bacterial cellulose. Adv Funct Mater 25:3560–3570. https://doi.org/10.1002/adfm.201500673
98. Song J, Jeon JH, Oh IK, Park KC (2011) Electro-active polymer actuator based on sulfonated polyimide with highly conductive silver electrodes via self-metallization. Macromol Rapid Commun. https://doi.org/10.1002/marc.201100318
99. Jung J-H, Jeon J-H, Sridhar V, Oh I-K (2011) Electro-active graphene–Nafion actuators. Carbon 49:1279–1289. https://doi.org/10.1016/j.carbon.2010.11.047
100. Mukai K et al (2009) Highly conductive sheets from millimeter-long single-walled carbon nanotubes and ionic liquids: application to fast-moving, low-voltage electromechanical actuators operable in air. Adv Mater 21:1582–1585. https://doi.org/10.1002/adma.200802817
101. Mahato M et al (2020) Covalent organic frameworks: sulfur- and nitrogen-rich porous π-conjugated COFs as stable electrode materials for electro-ionic soft actuators. Adv Funct Mater 30(46). https://doi.org/10.1002/adfm.202070300
102. Park JH, Lee SW, Song DS, Jho JY (2015) Highly enhanced force generation of ionic polymer-metal composite actuators via thickness manipulation. ACS Appl Mater Interfaces 7:16659–16667. https://doi.org/10.1021/acsami.5b04296

103. Tabassian R et al (2019) Graphene mesh for self-sensing ionic soft actuator inspired from mechanoreceptors in human body. Adv Sci 6:1901711. https://doi.org/10.1002/advs.201901711
104. Tabassian R et al (2019) Graphene mesh for self-sensing ionic soft actuator inspired from mechanoreceptors in human body. Adv Sci. https://doi.org/10.1002/advs.201901711
105. Panwar V, Cha K, Park JO, Park S (2012) High actuation response of PVDF/PVP/PSSA based ionic polymer metal composites actuator. Sens Actuators B Chem 161:460–470. https://doi.org/10.1016/j.snb.2011.10.062
106. Nguyen VH, Oh I-K (2021) Block copolymer electrolyte for ionic electroactive artificial muscles, vol 11587 SS. SPIE
107. Lin J, Liu Y, Zhang QM (2011) Charge dynamics and bending actuation in aquivion membrane swelled with ionic liquids. Polym (Guildf) 52:540–546. https://doi.org/10.1016/j.polymer.2010.11.030
108. Nemat-Nasser S, Wu Y (2003) Comparative experimental study of ionic polymer–metal composites with different backbone ionomers and in various cation forms. J Appl Phys 93:5255. https://doi.org/10.1063/1.1563300
109. Lee JW et al (2018) Styrenic block copolymer/sulfonated graphene oxide composite membranes for highly bendable ionic polymer actuators with large ion concentration gradient. Compos Sci Technol 163:63–70. https://doi.org/10.1016/j.compscitech.2018.05.002
110. Lee J-W, Hong SM, Kim J, Koo CM (2012) Novel sulfonated styrenic pentablock copolymer/silicate nanocomposite membranes with controlled ion channels and their IPMC transducers. Sens Actuators B Chem 162:369–376. https://doi.org/10.1016/j.snb.2011.12.105
111. Lee JW, Yu S, Hong SM, Koo CM (2013) High-strain air-working soft transducers produced from nanostructured block copolymer ionomer/silicate/ionic liquid nanocomposite membranes. J Mater Chem C 1:3784–3793. https://doi.org/10.1039/c3tc30414k
112. Nan M et al (2019) Ecofriendly high-performance ionic soft actuators based on graphene-mediated cellulose acetate. Sens Actuators B Chem 301:127127. https://doi.org/10.1016/j.snb.2019.127127
113. Zhao G, Yang J, Wang Y, Zhao H, Wang Z (2018) Preparation and electromechanical properties of the chitosan gel polymer actuator based on heat treating. Sens Actuators A 279:481–492. https://doi.org/10.1016/j.sna.2018.06.036
114. He QS, Yu M, Yang X, Kim KJ, Dai ZD (2015) An ionic electro-active actuator made with graphene film electrode, chitosan and ionic liquid. Smart Mater Struct 24:065026. https://doi.org/10.1088/0964-1726/24/6/065026
115. Goncalves R, Tozzi KA, Saccardo MC, Zuquello AG, Scuracchio CH (2020) Nafion-based ionomeric polymer/metal composites operating in the air: theoretical and electrochemical analysis. J Solid State Electr 24:1845–1856. https://doi.org/10.1007/s10008-020-04520-6
116. Imaizumi S, Kokubo H, Watanabe M (2012) Polymer actuators using ion-gel electrolytes prepared by self-assembly of ABA-triblock copolymers. Macromolecules 45:401–409. https://doi.org/10.1021/Ma2022138
117. Wu T et al (2012) RAFT synthesis of ABA triblock copolymers as ionic liquid-containing electroactive membranes. ACS Appl Mater Interfaces 4:6552–6559. https://doi.org/10.1021/am301662s
118. Green MD et al (2012) Synthesis of imidazolium ABA triblock copolymers for electromechanical transducers. Polymer 53:3677–3686. https://doi.org/10.1016/j.polymer.2012.06.023
119. He QS et al (2020) The highly stable air-operating ionic polymer metal composite actuator with consecutive channels and its potential application in soft gripper. Smart Mater Struct 29:045013. https://doi.org/10.1088/1361-665X/ab73e3
120. Tarokh A, Karan K, Ponnurangam S (2020) Atomistic MD study of Nafion dispersions: role of solvent and counterion in the aggregate structure, ionic clustering, and acid dissociation. Macromolecules 53:288–301. https://doi.org/10.1021/acs.macromol.9b01663
121. Welch C et al (2012) Nafion in dilute solvent systems: dispersion or solution? ACS Macro Lett 1:1403–1407. https://doi.org/10.1021/mz3005204

122. Shahinpoor M, Kim KJ (2001) Ionic polymer-metal composites: I. Fundamentals. Smart Mater Struct 10:819–833. https://doi.org/10.1088/0964-1726/10/4/327
123. Lu L, Liu J, Hu Y, Zhang Y, Chen W (2013) Graphene-stabilized silver nanoparticle electrochemical electrode for actuator design. Adv Mater 25:1270–1274. https://doi.org/10.1002/adma.201203655
124. Kim J, Tabassian R, Nguyen VH, Umrao S, Oh IK (2019) Crumpled quaternary nanoarchitecture of sulfur-doped nickel cobalt selenide directly grown on carbon cloth for making stronger ionic soft actuators. ACS Appl Mater Interfaces. https://doi.org/10.1021/acsami.9b12307
125. Park M, Kim J, Song H, Kim S, Jeon M (2018) Fast and stable ionic electroactive polymer actuators with PEDOT:PSS/(graphene(-)Ag-nanowires) nanocomposite electrodes. Sens (Basel) 18. https://doi.org/10.3390/s18093126
126. Geim AK, Novoselov KS (2007) The rise of graphene. Nat Mater 6:183–191. https://doi.org/10.1038/nmat1849
127. Wu ZS et al (2012) Graphene/metal oxide composite electrode materials for energy storage. Nano Energy 1:107–131. https://doi.org/10.1016/j.nanoen.2011.11.001
128. Liu M, Guo L, Jin S, Tan B (2019) Covalent triazine frameworks: synthesis and applications. J Mater Chem A 7:5153–5172. https://doi.org/10.1039/c8ta12442f
129. Bar-Cohen Y (2005) II ECCOMAS thematic conference on smart structures and materials
130. Bourzac K (2015) Spiders: web of intrigue. Nature 519:S4-6. https://doi.org/10.1038/519S4a
131. Gosline JM, Denny MW, Demont ME (1984) Spider silk as rubber. Nature 309:551–552. https://doi.org/10.1038/309551a0
132. Zhao Y, Li Y, Hien KTT, Mizutani G, Rutt HN (2019) Observation of spider silk by femtosecond pulse laser second harmonic generation microscopy. Surf Interface Anal 51:56–60. https://doi.org/10.1002/sia.6545
133. Feske S, Concepcion AR, Coetzee WA (2019) Eye on ion channels in immune cells. Sci Signal 12. https://doi.org/10.1126/scisignal.aaw8014
134. Nicoll RA (1988) The coupling of neurotransmitter receptors to ion channels in the brain. Science 241:545–551. https://doi.org/10.1126/science.2456612
135. Bartos DC, Grandi E, Ripplinger CM (2015) Ion channels in the heart. Compr Physiol 5:1423–1464. https://doi.org/10.1002/cphy.c140069
136. Semple JW, Italiano JE Jr, Freedman J (2011) Platelets and the immune continuum. Nat Rev Immunol 11:264–274. https://doi.org/10.1038/nri2956
137. Feinstein MB (1978) Calcium in drug action. In: Feinstein MB (ed) Chapter 9, pp 197–239
138. Zhang X, Cote RH (2005) CGMP signaling in vertebrate retinal photoreceptor cells. Front Biosci 10:1191–1204. https://doi.org/10.2741/1612
139. Schroeder TBH et al (2017) An electric-eel-inspired soft power source from stacked hydrogels. Nature 552:214–218. https://doi.org/10.1038/nature24670
140. Yang PP et al (2020) A biomimetic platelet based on assembling peptides initiates artificial coagulation. Sci Adv 6:eaaz4107. https://doi.org/10.1126/sciadv.aaz4107
141. Ahn T et al (2021) Hierarchical nature of nanoscale porosity in bone revealed by positron annihilation lifetime spectroscopy. ACS Nano 15:4321–4334. https://doi.org/10.1021/acsnano.0c07478
142. Buckwalter JA, Glimcher MJ, Cooper RR, Recker R (1995) Bone biology. J Bone Joint Surg 77:1256–1275. https://doi.org/10.2106/00004623-199508000-00019
143. Wegst UG, Bai H, Saiz E, Tomsia AP, Ritchie RO (2015) Bioinspired structural materials. Nat Mater 14:23–36. https://doi.org/10.1038/nmat4089
144. Weiner S, Wagner HD (1998) The material bone: structure mechanical function relations. Annu Rev Mater Sci 28:271–298. https://doi.org/10.1146/annurev.matsci.28.1.271
145. Grandfield K, Vuong V, Schwarcz HP (2018) Ultrastructure of bone: hierarchical features from nanometer to micrometer scale revealed in focused ion beam sections in the TEM. Calcif Tissue Int 103:606–616. https://doi.org/10.1007/s00223-018-0454-9
146. Zhao Y, Hien KTT, Mizutani G, Rutt HN (2017) Second-order nonlinear optical microscopy of spider silk. Appl Phys B-Lasers O 123:188. https://doi.org/10.1007/s00340-017-6766-z

147. Rising A, Johansson J (2015) Toward spinning artificial spider silk. Nat Chem Biol 11:309–315. https://doi.org/10.1038/nchembio.1789
148. Flood E, Boiteux C, Lev B, Vorobyov I, Allen TW (2019) Atomistic simulations of membrane ion channel conduction, gating, and modulation. Chem Rev 119:7737–7832. https://doi.org/10.1021/acs.chemrev.8b00630
149. Koyama M et al (2017) Bone-like crack resistance in hierarchical metastable nanolaminate steels. Science 355:1055–1057. https://doi.org/10.1126/science.aal2766
150. Cahn RW (2006) Learning from nature. Nature 444:425–426. https://doi.org/10.1038/444425b
151. Sanchez C, Arribart H, Guille MM (2005) Biomimetism and bioinspiration as tools for the design of innovative materials and systems. Nat Mater 4:277–288. https://doi.org/10.1038/nmat1339
152. Snell-Rood E (2016) Interdisciplinarity: bring biologists into biomimetics. Nature 529:277–278. https://doi.org/10.1038/529277a
153. Ortiz C, Boyce MC (2008) Materials science. Bioinspired structural materials. Science 319:1053–1054. https://doi.org/10.1126/science.1154295
154. Katiyar NK, Goel G, Hawi S, Goel S (2021) Nature-inspired materials: emerging trends and prospects. Npg Asia Mater 13:56. https://doi.org/10.1038/s41427-021-00322-y
155. Ganewatta MS, Wang Z, Tang C (2021) Chemical syntheses of bioinspired and biomimetic polymers toward biobased materials. Nat Rev Chem 5:753–772. https://doi.org/10.1038/s41570-021-00325-x
156. Di Vincenzo M et al (2021) Biomimetic artificial water channel membranes for enhanced desalination. Nat Nanotechnol 16:190–196. https://doi.org/10.1038/s41565-020-00796-x
157. Lee H, Lee BP, Messersmith PB (2007) A reversible wet/dry adhesive inspired by mussels and geckos. Nature 448:338–341. https://doi.org/10.1038/nature05968
158. Autumn K et al (2000) Adhesive force of a single gecko foot-hair. Nature 405:681–685. https://doi.org/10.1038/35015073
159. Geim AK et al (2003) Microfabricated adhesive mimicking gecko foot-hair. Nat Mater 2:461–463. https://doi.org/10.1038/nmat917
160. Lee LP, Szema R (2005) Inspirations from biological optics for advanced photonic systems. Science 310:1148–1150. https://doi.org/10.1126/science.1115248
161. Laschi C, Mazzolai B, Cianchetti M (2016) Soft robotics: technologies and systems pushing the boundaries of robot abilities. Sci Robot 1. https://doi.org/10.1126/scirobotics.aah3690
162. Bretscher MS (1973) Membrane structure: some general principles. Science 181:622–629. https://doi.org/10.1126/science.181.4100.622
163. Nguyen VH et al (2024) Functionally antagonistic polyelectrolyte for electro-ionic soft actuator. Nat Commun 15:435. https://doi.org/10.1038/s41467-024-44719-z
164. Sivaraj KK, Adams RH (2016) Blood vessel formation and function in bone. Development 143:2706–2715. https://doi.org/10.1242/dev.136861
165. Arfat Y, Rani A, Jingping W, Hocart CH (2020) Calcium homeostasis during hibernation and in mechanical environments disrupting calcium homeostasis. J Comp Physiol B 190:1–16. https://doi.org/10.1007/s00360-019-01255-3
166. Garai M et al (2024) Multifunctional and electronically conjugated triazine framework for superior electro-ionic artificial muscles. Adv Funct Mater. https://doi.org/10.1002/adfm.202406603

Chapter 2
Cell-Membrane-Inspired Block Isomer Electrolyte

2.1 Introduction to Polymer Electrolytes for Ionic Artificial Muscles

Compliant actuators are essential components in the development of soft robotics. Among the various types, ionic actuators stand out due to their promising properties, including large deformations at low operating voltages (under 5 V), low power consumption, and ease of control through the adjustment of voltage amplitude and frequency [1–8]. However, the current performance of ionic actuators remains inadequate for most soft robotics applications. Therefore, enhancing their performance is a vital step toward their practical implementation.

Ionic actuators typically consist of two electrodes encasing a polymer electrolyte membrane. Thus, material optimization is a key strategy for improving their performance. Although extensive research has been conducted on both electrolyte and electrode materials, the focus has predominantly been on electrodes, with significantly less attention given to electrolytes, with few examples listed in Fig. 2.1a [1, 2, 4–8, 11–15]. This is likely due to the challenge of developing electrolyte membranes that can achieve both high ionic conductivity and sufficient mechanical strength. Typically, increasing ionic conductivity by expanding conducting phases results in a compromise in mechanical strength due to the reduction of non-conducting domains, and vice versa. Additionally, there is a notable gap in effective techniques for forming continuous conducting phases in micro-thick films.

The unique amphiphilic structure and self-assembly properties of the cell (or plasma) membrane provide a compelling model for improving ionic conduction in micro-thick films. The cell membrane, which forms the outermost layer of animal cells and the second outermost layer of plant cells, is composed primarily of amphiphilic lipids, such as phospholipids. These materials feature a hydrophilic phosphate group (the head) and two hydrophobic hydrocarbon chains (the tails). This amphiphilic structure drives the self-assembly of phospholipids into a bilayer, with the hydrophilic heads facing outward and the hydrophobic tails oriented inward

V. H. Nguyen, *Bioinspired Multifunctional Nanomaterials for Ionic Artificial Muscles*, Springer Theses, https://doi.org/10.1007/978-3-031-78813-0_2

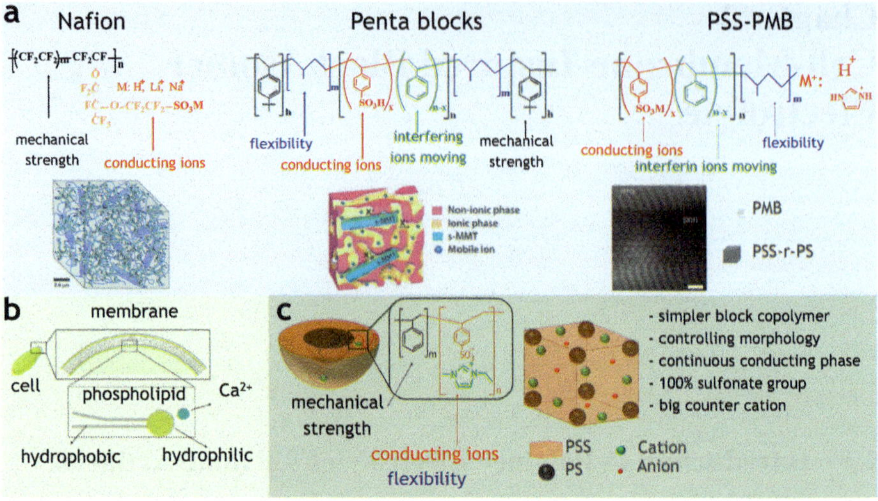

Fig. 2.1 Idea development: **a** previous polymer electrolyte, Nafion [9] from Ref. [9]. Reprinted with permission from the Royal Society of Chemistry, penta blocks [7] from Ref. [7]. Reprinted with permission from the Royal Society of Chemistry, and PSS-PMB [2] from Ref. [2]. Reprinted with permission from Springer Nature; **b** structure and composition of cell membrane; **c** cell-membrane-inspired PS-*b*-PSS-EMIm electrolyte [10] from Ref. [10]. Reprinted with permission from John Wiley and Sons

(Fig. 2.1b). Despite being only about five nanometers thick, this bilayer is robust enough to support cellular function by maintaining the separation between intracellular and extracellular environments and selectively transporting materials essential for cellular metabolism. The tails stabilize the membrane in aqueous environments, while the heads enable effective interaction with the surrounding environments. The self-assembly mechanism in phospholipids thus serves as an inspiration for designing polymer electrolytes that could self-assemble into electrolyte membranes with both high ionic conductivity and adequate mechanical strength (Fig. 2.1c).

Block isomers are particularly suitable for this approach because they can integrate an ionic block and a covalent block within a single macromolecule. These blocks can self-assemble into specific morphologies that promote both ionic conductivity and mechanical strength. Several commercial block isomers [10, 16–18] have been tested, and new block isomers [2, 19–23] have been synthesized to improve actuation performance. However, despite some progress, existing block isomers have not fully overcome the challenge of balancing ionic conductivity and mechanical strength. This difficulty may stem from the fact that while block isomers theoretically allow for the formation of continuous conducting phases needed for high ionic conductivity, experimental control over such self-assembly in micro-thick films for actuator applications remains challenging, as current techniques are primarily designed for nanoscale thin films.

This chapter presents an innovative, cell-membrane-inspired polystyrene-*b*-poly(1-ethyl-3-methylimidazolium-4-styrenesulfonate) (PS-*b*-PSS-EMIm) that successfully addressed the inverse relationship between ionic conductivity and mechanical strength in micro-thick electrolyte membranes (Fig. 2.1b and c) [10]. Analogous to the cell membrane, the PS non-conducting block provides necessary mechanical strength, similar to the role of the hydrophobic tail in phospholipid; the PSS conducting block, composed of styrenesulfonate units, facilitates ion movement in a manner akin to the hydrophilic head of phospholipid; and the EMIm mobile cation serves a function similar to that of calcium cations in the cell membrane. A carefully selected synthesis method was employed to promote the self-assembly of this block copolymer into micelles during polymerization. These micelles, featuring PS cores and PSS-EMIm shells, formed continuous PSS-EMIm conducting phases and distinct PS non-conducting phases in micro-thick membranes. As a result, the corresponding actuators demonstrated significantly improved performance.

2.2 Experiment and Characterization

2.2.1 Materials

The main chemicals used for this chapter include sodium 4-vinylbenzenesulfonate (SSNa, 90%), ethyl 2-bromopropionate (99%), potassium ethyl xanthogenate (96%), divinylbenzene (80%) from Sigma-Aldrich; 1-ethyl-3-methylimidazolium chloride (EMImCl, 98%) and 1-ethyl-3-methylimidazolium tetrafluoroborate ($EMImBF_4$, 98%) from Io.li.tec.; PP (CLEVIOS PH 1000) from Heraeus; styrene (93%) from Junsei; alumina (basic activated type) from Acros. All materials were used as received except for divinylbenzene and styrene, which were passed through alumina columns to remove their inhibitors before use. S-2-ethyl propionate-o-ethyl xanthate was synthesized according to the literature [24].

2.2.2 PS-b-PSS-EMIm Synthesis and Characterization

The SS-EMIm monomer was synthesized from SSNa and EMImCl in DMF (Fig. 2.2a). The product was extracted and characterized by proton nuclear magnetic resonance (^{1}H NMR). PS-*b*-PSS-EMIm core–shell particles were polymerized via the reversible addition-fragmentation chain-transfer (RAFT) technique initiated by ultraviolet irradiation (Fig. 2.2b). These products were tested by gel permeation chromatography (GPC), ^{1}H NMR, scanning and transmission electron microscopy (SEM, TEM).

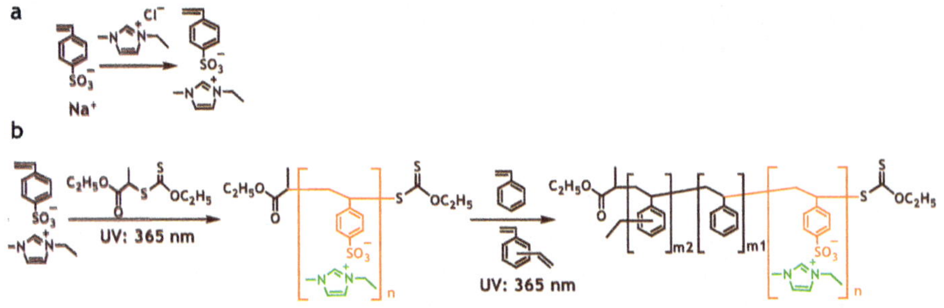

Fig. 2.2 Synthesis: **a** SS-EMIm monomer; **b** PS-*b*-PSS-EMIm block copolymer

2.2.3 *Fabrication and Investigation of Electrolyte Membranes and Actuators*

Electrolyte membranes were cast from mixtures of PS-*b*-PSS-EMIm emulsion and $EMImBF_4$ in Teflon molds at 50 °C, then intensely dried in vacuum at 80 °C (Fig. 2.3), and analyzed by differential scanning calorimetry (DSC), electrochemical impedance spectroscopy (EIS), and tensile testing.

Nitrogen and sulfur co-doped reduced graphene oxide prepared according the literature [5] was mixed with PP and cast on a Teflon mold at 80 °C, resulting in dry electrode films. The electrolyte membrane and electrode films were hot pressed together, forming actuators as illustrated in Fig. 2.3. These actuators were activated by direct current (DC) and alternating current (AC) voltages with different amplitudes

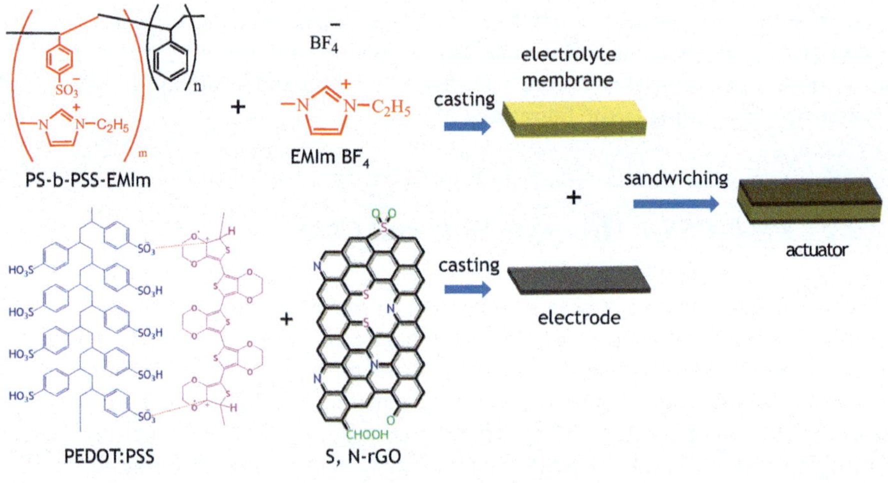

Fig. 2.3 Actuator fabrication [10] from Ref. [10]. Reprinted with permission from John Wiley and Sons

and frequencies. The resultant displacements and blocking forces were measured by laser sensors and load cells, respectively.

2.3 Results and Discussion

2.3.1 PS-b-PSS-EMIm Synthesis and Characterization

Proper procedures were established for synthesizing the SSEMIm monomer and PS-*b*-PSS-EMIm micelles. The SS-EMIm monomer was first prepared by a precipitation reaction between SSNa and EMImCl in DMF, where NaCl was formed as a by-product and precipitated out. This monomer was polymerized in water using the RAFT technique, with S-2-ethyl propionate-o-ethyl xanthate serving as the initiator, which was activated by ultraviolet irradiation, resulting in the formation of the PSS-EMIm conducting block. This living block was subsequently used in an emulsion polymerization process with styrene and divinyl benzene monomers, yielding micelles composed of PSS-EMIm conducting shells surrounding cross-linked PS non-conducting cores. Characterization of the monomer, conducting block, and core–shell micelles was carried out using ^{1}H NMR, GPC, and TEM, with results presented in Fig. 2.4.

The ^{1}H NMR spectrum clearly shows well-defined and integrated peaks corresponding to the EMIm monomer, as seen in Fig. 2.4a. The absence of impurity peaks indicates that the monomer was adequately purified and suitable for use in RAFT polymerization. After polymerization, the ^{1}H NMR spectra in Fig. 2.4a reveal the disappearance of the double bone peaks associated with the styrene sulfonate monomer, alongside the appearance of peaks corresponding to polystyrene sulfonate. This confirms the successful conversion of SS-EMIm monomer into the PSS-EMIm block. GPC analysis, depicted in Fig. 2.4b, indicates that the number-average molecular weight of the PSS-EMIm block was approximately 145,000 g/mol. Despite its high molecular weight and a relatively low transfer coefficient of 0.67 for the xanthate initiator [25], the block exhibited a low polydispersity index of 1.8, demonstrating that the initiator, polymerization method, and conditions were appropriate for preparing the PSS-EMIm block. Together, these GPC and ^{1}H NMR results confirm the successful synthesis of the SS-EMIm monomer and the PSS-EMIm conducting block.

The addition of styrene to the PSS-EMIm block resulted in the formation of PS-*b*-PSS-EMIm core–shell micelles, as evidenced by the data in Fig. 2.4c–h. To clearly observe the core–shell structure, a single particle was examined using high-angle annular dark-field (HAADF) imaging and elemental mapping in TEM. HAADF is particularly effective in detecting contrast between elements based on their atomic numbers, where higher atomic number elements scatter more electrons, resulting in brighter areas. In the case of PS-*b*-PSS-EMIm, the sulfur atoms in the shells have an atomic number more than 2.5 times that of the carbon atoms in the cores,

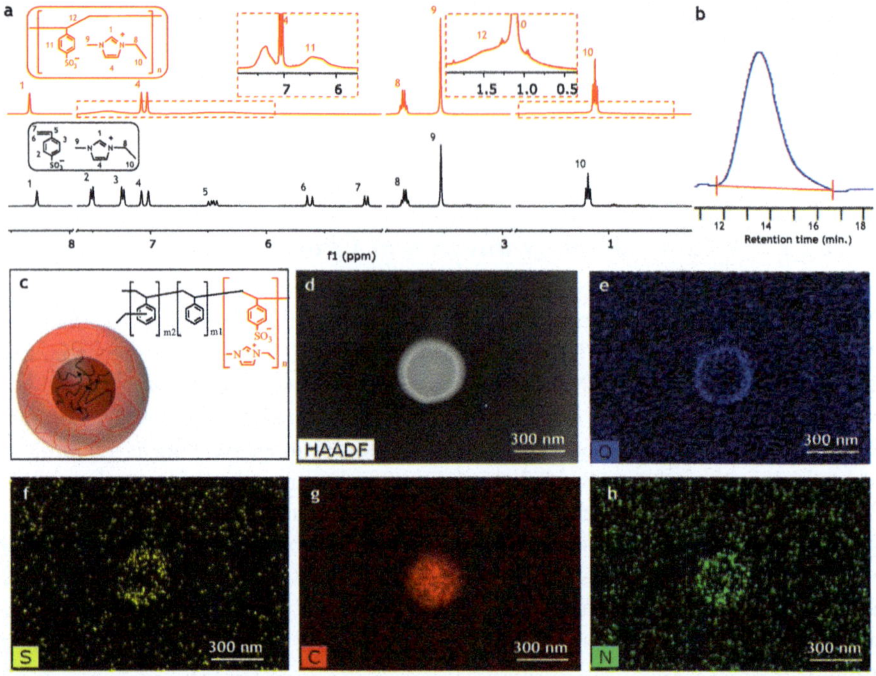

Fig. 2.4 Properies of PS-*b*-PSS-EMIm core–shell particles [10] from Ref. [10]. Reprinted with permission from John Wiley and Sons: **a** ^{1}H NMR and **b** GPC trace of PSS-EMIm; **c** illustration; **d** TEM; **e** oxygen, **f** sulfur, **g** carbon, **h** nitrogen elemental mapping images

providing sufficient contrast for clear visualization. As shown in Fig. 2.4d, the core–shell structure is evident, with the bright PSS-EMIm shell encasing the darker PS core. This contrast is attributable to the sulfur in the shell (atomic number 16) and the carbon in the core (atomic number 6). The HAADF image was further confirmed by elemental mapping, as illustrated in Fig. 2.4e–h, where oxygen, sulfur, and nitrogen were localized in the PSS-EMIm shell, while carbon was uniformly distributed across the entire particle.

2.3.2 *Properties of Electrolyte Membrane*

Membranes were cast from a mixture of PS-*b*-PSS-EMIm core–shell particles and $EMImBF_4$ ionic liquid. A large quantity of PS-*b*-PSS-EMIm core–shell particles was observed by SEM, and these membranes were subsequently characterized by tensile test, DSC, and EIS. The summarized results in Fig. 2.5 provide insights into the mechanical strength, phase separation, and ionic conductivity of the as-produced membranes.

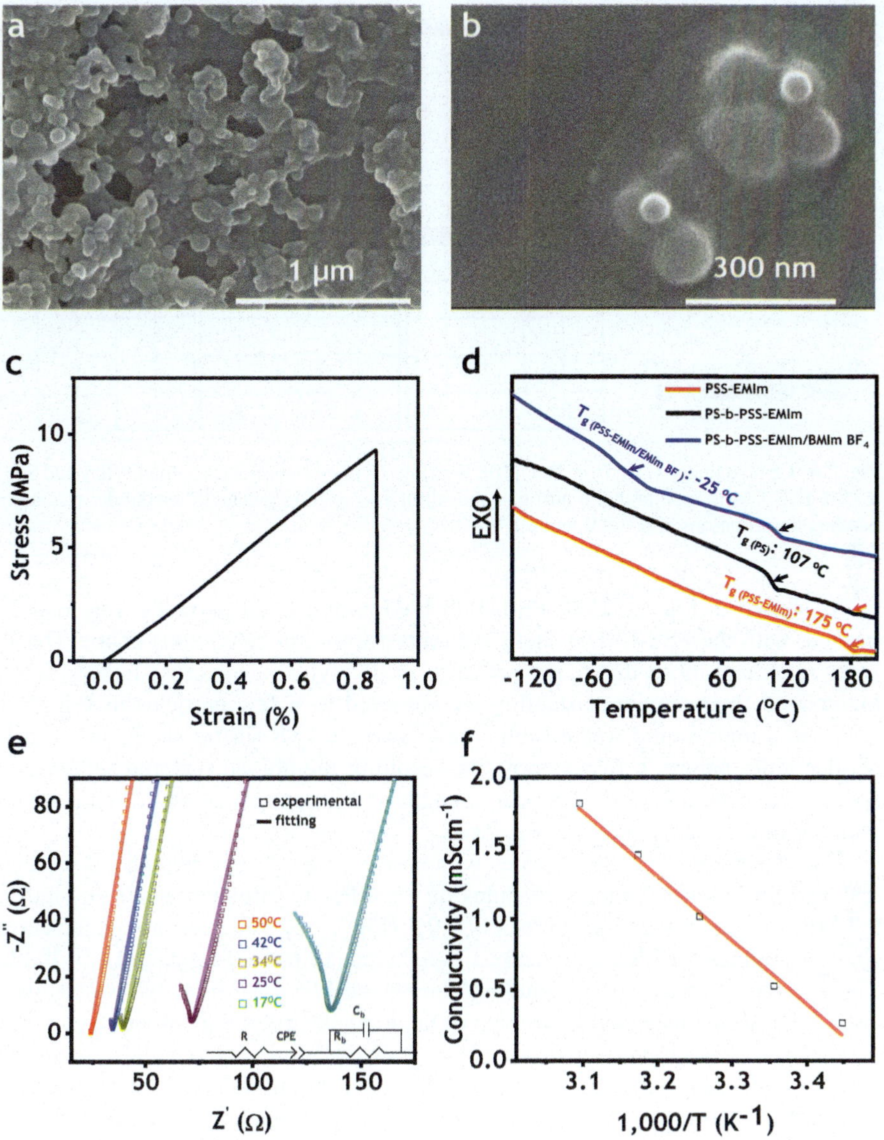

Fig. 2.5 Properties of PS-*b*-PSS-EMIm membrane [10] from Ref. [10]. Reprinted with permission from John Wiley and Sons: **a–b** field emission scanning electron microscopy (FE-SEM) images; **c** tensile test; **d** DSC curves; **e** EIS spectra; **f** ionic conductivity

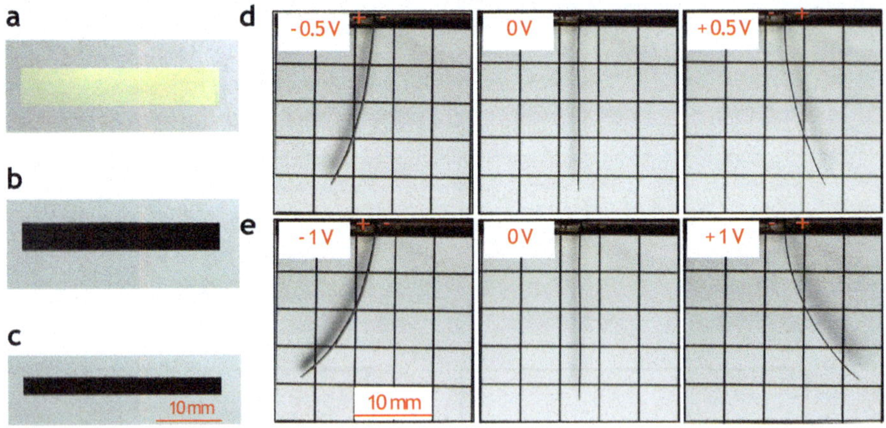

Fig. 2.6 Optical images [10] from Ref. [10]. Reprinted with permission from John Wiley and Sons: **a** PS-*b*-PSS-EMIm membrane; **b** nitrogen and sulfur co-doped graphene/PP electrode; **c** actuator; bending deformations at **d** 0.5 V and **e** 1 V

As shown in Fig. 2.5a, all PS-b-PSS-EMIm core–shell particles were densely packed, with the PSS-EMIm shells naturally forming a continuous phase. The PS cores remained intact due to the presence of divinyl benzene cross-linkers, which maintained their structural stability, as observed in a few particles in Fig. 2.5b. $EMImBF_4$ ionic liquid was miscible only with the PSS-EMIm shells due to their similar ionic nature. Consequently, the resulting membrane featured separate PS spheres contributing to mechanical strength and a continuous PSS-EMIm matrix facilitating ionic conductivity (Fig. 2.6a).

The PS-*b*-PSS-EMIm membrane, without ionic liquids, exhibited good mechanical strength, with a Young's modulus of 1.1 GPa, an ultimate tensile strength of 9.3 MPa, and an elongation at break of 0.9% (Fig. 2.5c). For comparison, the corresponding values for PS are typically 1.6–3.4 GPa, 17.6–53.0 MPa, and 1.0–70.0%. These results suggest strong bonding among the PSS-EMIm shells, which likely contributes to the mechanical strength of the final electrolyte membranes.

For application in ionic actuators, PS-*b*-PSS-EMIm was mixed with $EMImBF_4$, which served dual purposes: providing additional mobile ions and plasticizing the PSS-EMIm matrix to enhance its flexibility for ion conduction. The DSC results in Fig. 2.5d indicate that the complete incorporation of $EMImBF_4$ with the PSS-EMIm shells resulted in a single, clear glass transition temperature (T_g), which sharply decreased from 175 to −25 °C. This low T_g value indicates that the PSS-EMIm phase became highly flexible, significantly enhancing ion mobility. Simultaneously, the T_g values of the PS block remained unchanged at 107 °C, suggesting that the PS cores did not mix with $EMImBF_4$. This T_g value is slightly higher than that of conventional PS, likely due to the introduction of divinyl benzene cross-linkers during synthesis, which helped preserve the mechanical integrity of the PS cores during mixing with the ionic liquid, thereby contributing to the robustness of the free-standing electrolyte membranes. In contrast, PSS-EMIm without the PS block formed gels when mixed with the same amount of $EMImBF_4$, highlighting the importance of PS cores for the mechanical strength of both the electrolyte membranes and the actuators.

The key innovation of this research lies in the unique core–shell structure that enabled the fabrication of thick electrolyte membranes (80 μm) with a continuous conducting phase. Ionic conductivity was measured by EIS over a broad temperature range (17–50 °C), as shown in Fig. 2.5e and f. The membrane demonstrated high ionic conductivity of 0.51 mS/cm at room temperature, which increased with temperature following the Arrhenius law.

The results presented in Figs. 2.4 and 2.5 demonstrate that this block copolymer, with unique core–shell structure, successfully formed a continuous conducting phase alongside separate non-conducting domains. This effectively overcame the inverse relationship between ionic conductivity and mechanical strength. Consequently, the electrolyte membrane achieved both high ionic conductivity, thanks to the continuous PSS-HMIm conducting phase, and good mechanical strength, due to the separate cross-linked PS non-conducting domains, making it suitable for ionic actuator applications.

2.3.3 Actuation Performance

The optical images of the polymer membrane, electrode, actuator, and its bending deformations are depicted in Fig. 2.6. Figure 2.6a–c illustrate smooth and uniform surfaces of the electrolyte membrane, electrode film, and actuator after hot pressing. Notably, the presence of the same sulfonate groups facilitated strong bonding between the electrolyte membrane and electrodes. Figure 2.6d and e display the activated actuator, which exhibited large and symmetrical tip-to-tip bending deformations of approximately 20 mm and 40 mm at 0.5 and 1.0 V, respectively.

To further assess the quality of the developed membrane, the actuator was extensively tested under various conditions, including DC voltages ranging from 0.1 to 2.0 V and AC input voltages from 0.05 to 1.00 V at frequencies ranging from 0.1 to 5.0 Hz. The resulting data were used to evaluate actuation performance in term of deformation, rise time, durability, and blocking force, as summarized in Figs. 2.7 and 2.8.

As shown in Fig. 2.7a and b, excitations of 0.5 V and 0.1 Hz caused the actuator, with a free length of 20 mm, to exhibit significantly bending, with displacements of 8.2 mm (0.37% strain) and 6.8 mm (0.32% strain) for square and harmonic voltages, respectively. The square voltage generated faster and larger bending than the harmonic one, due to its quicker attainment of maximum amplitude.

The DC response results in Fig. 2.7c indicate that the actuator had a large and fast response, with a bending displacement of approximately 4.1 mm (0.21% strain) and a rise time af about five seconds at an input of 0.5 V. Furthermore, the deformation increased linearly with DC inputs ranging from 0.1 to 0.5 V, and these DC deformations remained consistent without back relaxation over more than 220 s.

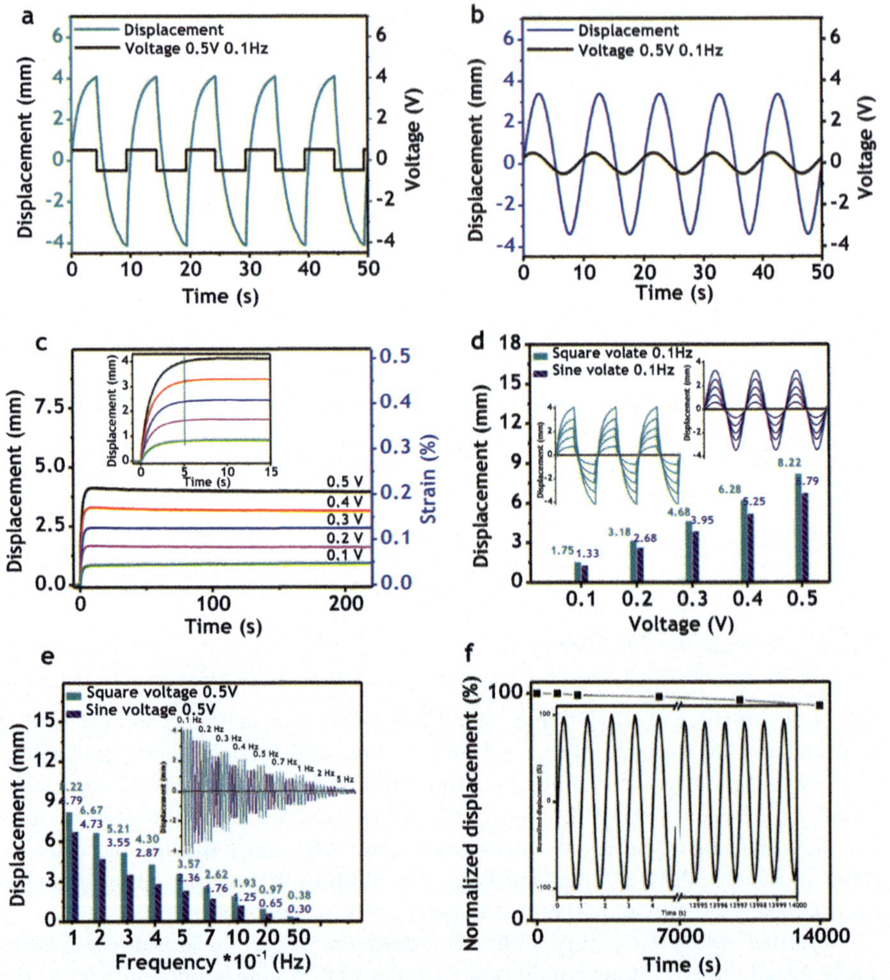

Fig. 2.7 Actuation performance [10] from Ref. [10]. Reprinted with permission from John Wiley and Sons: bending deformations (0.5 V and 0.1 Hz) of **a** square and **b** sine voltages; **c** deformations under DC voltages; deformations according to **d** voltages and **e** frequencies; **f** durability

Figure 2.7d and e demonstrate that the actuator exhibited significant bending across a wide range of low voltages (0.1–0.5 V) and frequencies (0.1–5.0 Hz), highlighting its potential for low-power-driven applications. The observed deformations increased with voltage amplitude and decreased with rising frequency. This behavior can be attributed to the fact that higher voltages generate stronger ionic currents within the electrolyte membrane, while faster frequencies provide less time for ions to reach the electrodes.

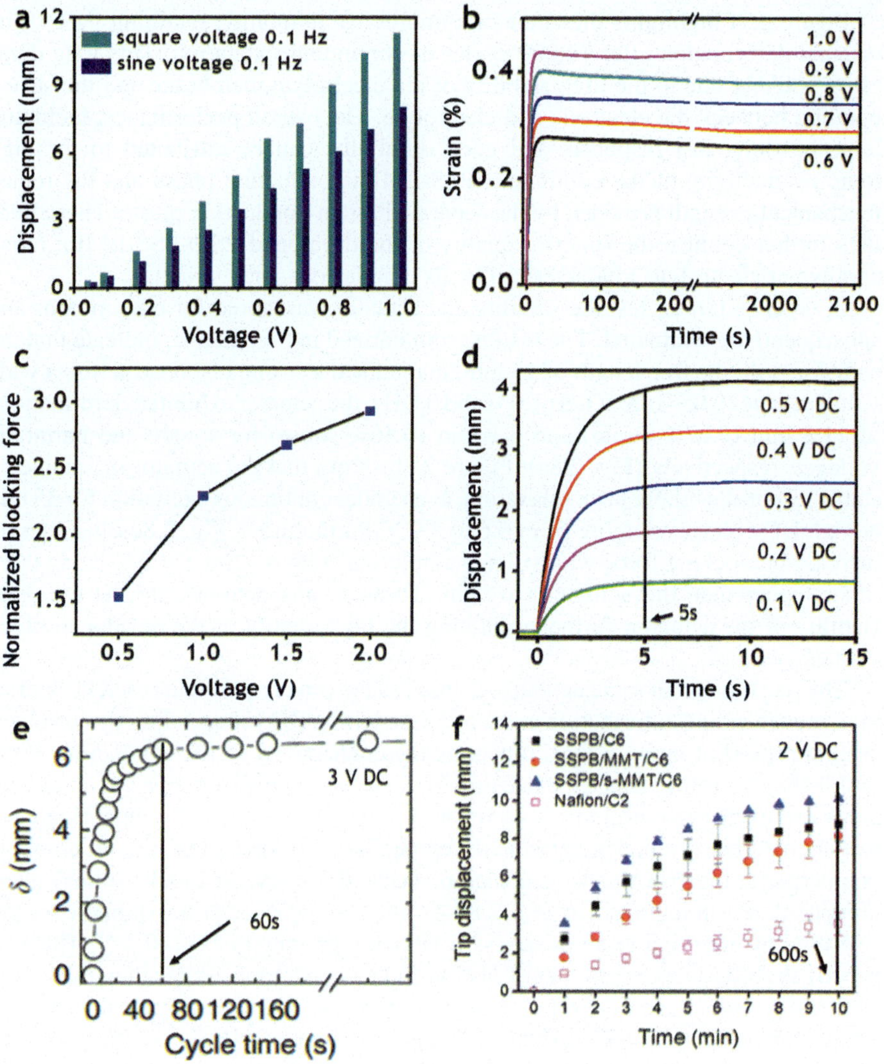

Fig. 2.8 Performance after nine months and rise time comparison: **a** displacement; **b** strain; **c** mormalized blocking force comparing to the weight of the actuator; **d** rise time [10] from Ref. [10]. Reprinted with permission from John Wiley and Sons, in comparison with **e** PSS-*b*-PMB [2] from Ref. [2]. Reprinted with permission from Springer Nature and **f** penta-block from Ref. [7]. Reprinted with permission from the Royal Society of Chemistry

Figure 2.7f highlights the actuator's impressive durability, retaining 94% of its original deformation after 14,000 cycles of continuous excitation. This long durability is likely due to the high stability of the electrolyte membrane and the strong bonding between the electrode and electrolyte. The overall performance, including large bending, fast response, and good durability, can be attributed to the high ionic conductivity of the continuous PSS-EMIm conducting phase and the robust mechanical strength provided by the separate PS non-conducting phase. These findings further confirm the superior quality of the developed PS-*b*-PSS-EMIm block copolymer electrolyte, characterized by its unique core–shell structure.

In order to further test the stability, the actuator was stored for nine months and subsequently re-evaluated. The results, summarized in Fig. 2.8a–c, indicate that the actuator, with the free length of 15 mm, maintained a strong response across a wide voltage range (0.05–1.00 V). At 1.0 V and 0.1 Hz, the actuator exhibited deformations of 11.3 mm (0.67% strain) and 8.1 mm (0.56% strain) for square and harmonic voltages, respectively (Fig. 2.8a). Figure 2.8b shows that the actuator maintained a stable response without back relaxation, even under continuous excitation for 35 min at high DC voltages ranging from 0.6 to 1.0 V. Furthermore, Fig. 2.8c illustrates an increase in blocking force with voltage amplitude, with the force at 2 V being three times higher than the actuator's weight. These results demonstrate the excellent stability of the actuator, further validating the high quality of the developed block isomer.

The rise time of an ionic actuator, defined as the time required to reach 90% of its maximum bending deformation under a given exciting DC voltage, directly correlates with the speed of moving ions, reflecting the connectivity of the conducting phase within the electrolyte. Since the rise time is relatively independent of actuator dimensions and exciting conditions, it serves as a reliable parameter for comparing electrolyte materials. To benchmark the developed block isomer against other common counterparts, the rise times of actuators made from several block isomers were compared, as summarized in Fig. 2.8d–f. The PS-*b*-PSS-EMIm exhibited a short rise time of only five seconds across a DC voltage range of 0.1–0.5 V. This rise time is slightly shorter than that of Nafion (6 s), twelve times faster than that of the PSS-PMB diblock isomer (60 s), and an impressive 120 times faster than that of the pentablock isomer (600 s) [2, 6, 7]. This comparison clearly demonstrates the superior quality of the PS-*b*-PSS-EMIm electrolyte membranes, attributed to their continuous conducting phase.

2.4 Conclusion

This chapter has presented the development of unique PS-*b*-PSS-EMIm core–shell particles, inspired by the assembly of amphiphilic phospholipid into cell membranes. The successful synthesis of these core–shell particles via RAFT polymerization was confirmed by ^{1}H NMR, GPC, SEM, and TEM analyses. The high quality of the thick membrane derived from these core–shell particles was further verified by SEM,

DSC, EIS, and tensile testing. Importantly, the PS-*b*-PSS-EMIm core–shell particles offer an effective method for controlling the morphology of the block copolymer in micro-thick electrolyte membranes, successfully addressing the inverse relationship between ionic conductivity and mechanical strength. The continuous conducting phase of PSS-EMIm, exclusively plasticized with $EMImBF_4$ ionic liquid, enabled high ionic conductivity, while the separate cross-linked PS spheres significantly enhanced mechanical strength. Consequently, the resulting actuator demonstrated exceptional performance across a wide range of conditions. Notably, this block copolymer electrolyte achieved a significant reduction in rise time compared to Nafion and other reported block copolymers.

References

1. Jo C, Pugal D, Oh I-K, Kim KJ, Asaka K (2013) Recent advances in ionic polymer–metal composite actuators and their modeling and applications. Prog Polym Sci 38:1037–1066. https://doi.org/10.1016/j.progpolymsci.2013.04.003
2. Kim O, Shin TJ, Park MJ (2013) Fast low-voltage electroactive actuators using nanostructured polymer electrolytes. Nat Commun 4:2208. https://doi.org/10.1038/ncomms3208
3. Bar-Cohen Y (2004) Electroactive polymer (EAP) actuators as artificial muscles: reality, potential, and challenges
4. Mukai K et al (2009) Highly conductive sheets from millimeter-long single-walled carbon nanotubes and ionic liquids: application to fast-moving, low-voltage electromechanical actuators operable in air. Adv Mater 21:1582-+. https://doi.org/10.1002/adma.200802817
5. Kotal M, Kim J, Kim KJ, Oh IK (2016) Sulfur and nitrogen co-doped graphene electrodes for high-performance ionic artificial muscles. Adv Mater 28:1610–1615. https://doi.org/10.1002/adma.201505243
6. Tabassian R, Kim J, Nguyen VH, Kotal M, Oh I-K (2018) Functionally antagonistic hybrid electrode with hollow tubular graphene mesh and nitrogen-doped crumpled graphene for high-performance ionic soft actuators. Adv Funct Mater 28:1705714. https://doi.org/10.1002/adfm.201705714
7. Lee JW, Yu S, Hong SM, Koo CM (2013) High-strain air-working soft transducers produced from nanostructured block copolymer ionomer/silicate/ionic liquid nanocomposite membranes. J Mater Chem C 1:3784–3793. https://doi.org/10.1039/c3tc30414k
8. Kim J, Jeon JH, Kim HJ, Lim H, Oh IK (2014) Durable and water-floatable ionic polymer actuator with hydrophobic and asymmetrically laser-scribed reduced graphene oxide paper electrodes. ACS Nano 8:2986–2997. https://doi.org/10.1021/nn500283q
9. Elliott JA, Wu DS, Paddison SJ, Moore RB (2011) A unified morphological description of Nafion membranes from SAXS and mesoscale simulations. Soft Matter 7:6820–6827. https://doi.org/10.1039/c1sm00002k
10. Nguyen VH et al (2019) Electroactive artificial muscles based on functionally antagonistic core-shell polymer electrolyte derived from PS-*b*-PSS block copolymer. Adv Sci (Weinh) 6:1801196. https://doi.org/10.1002/advs.201801196
11. Kong L, Chen W (2014) Carbon nanotube and graphene-based bioinspired electrochemical actuators. Adv Mater 26:1025–1043. https://doi.org/10.1002/adma.201303432
12. Vidal F, Plesse C, Teyssié D, Chevrot C (2004) Long-life air working conducting semi-IPN/ionic liquid based actuator. Synth Met 142:287–291. https://doi.org/10.1016/j.synthmet.2003.10.005
13. Lu C et al (2018) High-performance graphdiyne-based electrochemical actuators. Nat Commun 9:752. https://doi.org/10.1038/s41467-018-03095-1

14. Wu G et al (2015) Graphitic carbon nitride nanosheet electrode-based high-performance ionic actuator. Nat Commun 6:7258. https://doi.org/10.1038/ncomms8258
15. Fukushima T, Asaka K, Kosaka A, Aida T (2005) Fully plastic actuator through layer-by-layer casting with ionic-liquid-based bucky gel. Angew Chem Int Ed Engl 44:2410–2413. https://doi.org/10.1002/anie.200462318
16. Nguyen VH (2018) Electromechanical soft actuators based on polystyrene-b-poly (1-alkyl-3-methylimidazolium p-styrenesulfonate) block copolymer. Master thesis, Korea Advanced Institute of Science and Technology
17. Wang XL, Oh IK, Lu J, Ju JH, Lee S (2007) Biomimetic electro-active polymer based on sulfonated poly (styrene-b-ethylene-co-butylene-b-styrene). Mater Lett 61:5117–5120. https://doi.org/10.1016/j.matlet.2007.04.004
18. Vargantwar PH, Shankar R, Krishnan AS, Ghosh TK, Spontak RJ (2011) Exceptional versatility of solvated block copolymer/ionomer networks as electroactive polymers. Soft Matter 7:1651. https://doi.org/10.1039/c0sm01210f
19. Imaizumi S, Kokubo H, Watanabe M (2012) Polymer actuators using ion-gel electrolytes prepared by self-assembly of ABA-triblock copolymers. Macromolecules 45:401–409. https://doi.org/10.1021/Ma2022138
20. Wu T et al (2012) RAFT synthesis of ABA triblock copolymers as ionic liquid-containing electroactive membranes. ACS Appl Mater Interfaces 4:6552–6559. https://doi.org/10.1021/am301662s
21. Green MD et al (2012) Synthesis of imidazolium ABA triblock copolymers for electromechanical transducers. Polymer 53:3677–3686. https://doi.org/10.1016/j.polymer.2012.06.023
22. Kim O, Kim H, Choi UH, Park MJ (2016) One-volt-driven superfast polymer actuators based on single-ion conductors. Nat Commun 7:13576. https://doi.org/10.1038/ncomms13576
23. Kim O, Kim SY, Park B, Hwang W, Park MJ (2014) Factors affecting electromechanical properties of ionic polymer actuators based on ionic liquid-containing sulfonated block copolymers. Macromolecules 47:4357–4368. https://doi.org/10.1021/ma500869h
24. Destarac M, Brochon C, Catala J-M, Wilczewska A, Zard SZ (2002) Macromolecular design via the interchange of xanthates (MADIX): polymerization of styrene with o-ethyl xanthates as controlling agents. Macromol Chem Phys 203:2281–2289. https://doi.org/10.1002/macp.200290002
25. Moad G, Rizzardo E, Thang SH (2008) Radical addition-fragmentation chemistry in polymer synthesis. Polymer 49:1079–1131. https://doi.org/10.1016/j.polymer.2007.11.020

Chapter 3
Hemostasis-Inspired Electrolyte Membrane

3.1 Introduction to Electrolyte Membrane

Ionic electroactive actuators have shown significant potentials as a artificial muscle for soft robotics due to their large and rapid bending deformation, low power consumption, and safety when stimulated by voltages under 5 V [1–4]. These actuators can be easily integrated into control circuits [3, 5], and are capable of self-sensing [4, 6]. Their actuation mechanism is based on ion currents within an electrolyte membrane driven by the electrical field between two coated electrodes [7, 8]. Consequently, their performance is largely determined by the current density in the electrolyte membranes, which depends on the speed and quantity of ion movement. This is a function of both the frictional force (electrical resistivity) within the electrolyte membrane and the electrical force generated by the charged electrodes. Therefore, improving actuation performance can be achieved through advancements in electrolyte and electrode materials. While research on electrolytes has primarly focused on reducing the frictional force by lowering electrical resistivity through continuous conducting channels [5, 9], studies of electrodes have aimed to enhance the driving force by increasing electrical conductivity and capacitance [3, 8]. Unfortunately, polymer electrolytes have received less attention compared to electrode materials. This may be due to the challenge associated with producing membranes that exhibit both high ionic conductivity and good mechanical strength [5, 9–11]. Consequently, several polymer electrolytes have been reused, with Nafion being the most prominent [4, 7, 10–12].

Nafion, a random copolymer, consists of two immiscible monomer units: irregular hydrophilic side chains terminated with sulfonate groups and a hydrophobic tetrafluoroethylene backbone. The hydrophilic side chains aggregate into separate ionic clusters, resulting in poor ionic conductivity, while the hydrophobic backbone forms a continuous non-conducting matrix that provides mechanical strength [7, 13]. The random distribution of these side chains presents significant challenges in creating continuous conducting channels, particularly in thick films [5].

V. H. Nguyen, *Bioinspired Multifunctional Nanomaterials for Ionic Artificial Muscles*, Springer Theses, https://doi.org/10.1007/978-3-031-78813-0_3

As a result, there has been limited focus on controlling the morphology of this polymer in membranes. Typically, Nafion is dissolved in a good solvent such as N,N-dimethylacetamide (DMAc), cast into films, and subsequently applied to actuators [12], often without considering the interaction between Nafion molecules and the solvent, which can influence polymer configurations and, consequently, membrane properties. Recent theoretical [14, 15] and experimental [16, 17] studies have demonstrated that Nafion molecules adopt different conformations depending on the polarity of solvents (Fig. 3.1a). This insight offers a method for controlling the molecular configurations that promote the desired phase distribution in membranes. Specifically, Nafion micelles with sulfonate groups on their surfaces could form continuous conducting channels with low frictional force (electrical resistivity) and well-connected non-conducting domains that provide mechanical robustness.

Hemostasis, depicted in Fig. 3.1b, offers inspiration for developing electrolyte membranes with both high ionic conductivity and good mechanical strength from Nafion. Under normal conditions, platelets are covered by phospholipid membranes, with some calcium ions (Ca^{2+}) binding to their surfaces. During primary hemostasis, platelets are activated and acquire additional Ca^{2+} ions from blood plasma. The high concentration of Ca^{2+} ion disrupts cell membranes, transforming platelets into primary clots that cover damaged vessel areas. In secondary hemostasis, the abundant Ca^{2+} ions facilitate the production of crosslinked fibrins that reinforce the clots [19, 20]. This hemostasis process can be mimicked to fabricate suitable polyelectrolyte membranes from Nafion for use in ionic actuators. First, the amphiphilic nature of phospholipids suggests that Nafion, with hydrophilic side chains and hydrophobic backbone, is a good ionomer for conducting ions and providing mechanical strength. Second, the assembly of phospholipids into bilayers offers a strategy for controlling the morphology of Nafion at the molecular level through micellar formation. Finally, the formation of strong clots during hemostasis suggests a method for creating robust membranes from Nafion micelles.

This chapter presents a novel hemostasis-inspired procedure for producing Nafion membranes with high ionic conductivity and good mechanical strength, as described in Fig. 3.1c. First, mimicked the assembly of phospholipids into bilayer membranes, Nafion molecules were assembled into micelles in a water-DMAc mixture. Second, resembling primary hemostasis, the aggregation of these micelles occurred with the addition of $EMImBF_4$ ionic liquid. Third, analogous to secondary hemostasis, the interconnection of these micelles was formed during membrane casting at high temperature, followed by solvent evaporation. This procedure ultimately produced Nafion membranes with a continuous conducting phase for high ionic conductivity and well-connected non-conducting domains for mechanical strength. As a result, these membranes achieved over 35 times greater actuation bending, making them suitable for use in two applications: inchworm-inspired soft robots and dynamic tensegrity systems.

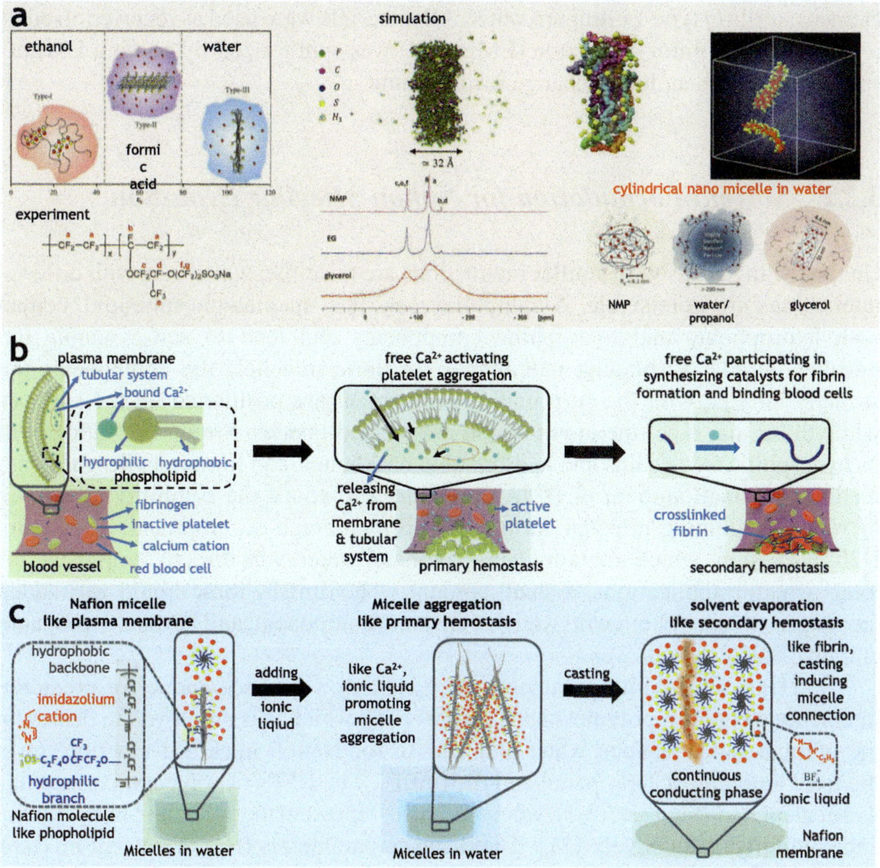

Fig. 3.1 Idea development: **a** dependence of Nafion molecule configuration on solvents [14–16] from Ref. [14]. Reprinted with permission from ACS Publications, from Ref. [15]. Reprinted with permission from John Wiley and Sons, from Ref. [16]. Reprinted with permission from ACS Publications; **b** platelet and hemostasis; **c** procedure for hemostasis-inspired membrane [18] from Ref. [18]. Reprinted with permission from Springer Nature

3.2 Experiment and Characterization

3.2.1 Materials

The main materials used in this research include D2021 Nafion solution from DuPont; 1-ethyl-3-methylimidazolium tetrafluoroborate (EMImBF$_4$, 98%) and 1-ethyl-3-methylimidazolium chloride (EMImCl, 98%) from Io.li.tec.; anhydrous N,N-dimethylacetamide (DMAc, 99,8%), anhydrous dimethylsulfoxide (DMSO, 99,9%), and Amberlite IRN78 ion-exchange resin from Sigma-Aldrich; poly(3,4-ethylenedioxythiophene)-poly(styrenesulfonic acid) (PP, CLEVIOS PH 1000) from

Heraeus; milli-Q type 1 ultrapure water. All materials were used as received. 1-ethyl-3-methylimidazolium hydroxide (EMImOH) was synthesized by passing EMImCl solutions through an ion-exchange resin column.

3.2.2 *Solvent Formulation for Nafion Micellar Evolution*

Generally, materials with similar interactions are miscible, while thos with differing interactions are immiscible. Amphiphilic materials, such as phospholipid, contain both hydrophobic and hydrophilic components and tend to self-assemble into micelles, similar to platelet membranes. In these micelles, the components that strongly interact with the surrounding enviroment are positioned on the surfaces, while those that repel the environment are located insides. Similarly, Nafion, with its hydrophilic side chains and hydrophobic backbones, can form micelles in a water medium. A small amount of DMAc was added to adjust the solubility parameters of water, facilitating micellar formation and membrane development. The Nafion D2021 solution, which contains 1-propanol and water, was directly employed. For ionic actuator applications, a small amount of $EMImBF_4$ ionic liquid was added, creating a solvent system with water as the main component, and DMAc, 1-propanol, and $EMImBF_4$ as additives.

The Hansen solubility parameter (HSP) is a powerful technique for predicting interactions between polymers and solvents [21], which was employed to determine the optimal ratio between water and DMAc for Nafion micelle formation, given the fixed amounts of 1-propanol and $EMImBF_4$. The HSPs of solvent mixtures (δ_{mj}) were calculated using Eq. (3.1), where Φ_i and δ_i represent the solvent volume fraction and solubility, respectively [21]. The distance parameters (R_a) of solvent mixtures relative to Nafion were also calculated using Eq. (3.2) [21]. Generally, if R_a is less than R_0 (the interaction radius of a polymer), the solvent mixture can dissolve the corresponding polymer. Since no reports on Nafion's R_0 exist in the literature, the calculated R_a values were compared to those of N-methyl-2-pyrrolidone (NMP, a good solvent for Nafion, $R_a = 2.69\ MPa^{1/2}$) and a water/2-propanol (1:1 by volume) mixture, which is a poor solvent system for Nafion ($R_a = 20.09\ MPa^{1/2}$) [16].

$$\delta_{mj} = \sum \Phi_i \delta_i \tag{3.1}$$

$$R_a^2 = 4\left(\delta_{d-solvent} - \delta_{d-polymer}\right)^2 + \left(\delta_{p-solvent} - \delta_{p-polymer}\right)^2 + \left(\delta_{h-solvent} - \delta_{h-polymer}\right)^2 \tag{3.2}$$

The HSP and R_a calculations suggested several water-to-DMAc ratios, which guided experimental investigations. This led to the identification of an appropriate solvent system for developing Nafion micelles and the hemostasis-inspired membrane (H-membrane). In addition, a conventional solvent system containing primarily DMAc and a small amount of $EMImBF_4$ was also analyzed (C-membrane).

Table 3.1 Solvent systems for preparing Nafion dispersions [18] from Ref. [18]. Reprinted with permission from Springer Nature

Solvent system	Solvent component [g]				Nafion [g]
	Water	1-Propanol	DMAc	$EMImBF_4$	
H-membrane	18.00	0.32	1.50	0.22	0.14
C-membrane	0.00	0.00	20.00	0.12	0.20

Table 3.2 Hansen solubility parameters and distance parameters of Nafion and solvents [18] from Ref. [18]. Reprinted with permission from Springer Nature

Material	Hansen solubility parameters [$MPa^{1/2}$]				Distance parameter, R_a [$MPa^{1/2}$]
	δ_t	δ_d	δ_p	δ_h	
Nafion [22]	23.5	17.4	12.5	9.6	$2.69 < R_0 < 20.09$
H-membrane	45.02	15.58	15.39	38.95	29.72
C-membrane	22.82	16.81	11.52	10.21	1.66
Water [21]	47.8	15.6	16.0	42.3	33.08
1-Propanol [21]	24.5	16.0	6.8	17.4	10.06
2-Propanol [21]	23.5	15.8	6.1	16.4	9.87
$EMImBF_4$ [23]	26.2	17.9	14.8	12.2	3.61
DMAc [21]	22.8	16.8	11.5	10.2	1.67
NMP [16, 21]	22.9	18.0	12.3	7.2	2.69
Water/2-propanol [16, 21]	35.7	15.7	11.1	29.6	20.09

All calculated values are summarized in Tables 3.1 and 3.2. It is important to note that the H-membrane mixture, with an R_a 1.5 times higher than that of the water/2-propanol mixture [16], is a poor solvent system that induces Nafion self-assembly into micelles. In contrast, the C-membrane blend, with an R_a 1.5 times smaller than that of NMP, is a good solvent combination that disperses Nafion effectively [16].

3.2.3 *Preparation and Investigation of Nafion Micelles*

Nafion micelles were prepared by diluting Nafion D2021 solution with water, and their micellar formation was confirmed using cryogenic transmission electron microscopy (cryo-TEM). The Nafion solution (0.70 g, stirred at 500 rpm and 90 °C) was diluted by slowly adding water (20.00 g), resulting in a dispersion that was maintained under the same conditions for 12 h before being allowed to cool naturally to 25 °C. The mixture was then left undisturbed for at least three days to ensure complete micelle evolution. The sulfonate acid groups of micelles were neutralized using caesium hydroxide, which also functioned as a staining agent. Finally, the stained micelles were deposited on a grid by a vitrobot and observed using a cryo field emission TEM (200 kV, Glacios, Thermo Fisher Scientific).

Next, the interaction between Nafion micelles with $EMImBF_4$ ionic liquid was examined using two techniques: static laser light scattering and dynamic light scattering (DLS) for micellar size, and zeta potential measurement for micellar surface charge. Because the addition of $EMImBF_4$ ionic liquid to the previous micellar dispersion produced a weak gel unsuitable for these characterizations, a more dilute dispersion was prepared. The Nafion solution (0.70 g, stirred at 500 rpm and 90 °C) was slowly mixed with DMAc (0.75 g) and water (48.00 g). After micelle evolution, this dispersion was neutralized with EMImOH (2.50 g solution in water) and mixed with $EMImBF_4$ (0.20 g in 0.75 g DMAc and 3.05 g water). A similar sample without ionic liquid was also prepared to assess the effect of $EMImBF_4$ on Nafion micelles. For comparison with conventional DMAc, two additional samples were prepared following the same procedures but using DMAc instead of water.

3.2.4 Fabrication and Investigation of Nafion Membranes

The procedues for Nafion membrane preparation are illustrated in Fig. 3.2. For the H-membrane (Fig. 3.2a), 2.50 g EMImOH solution in water and $EMImBF_4$ (0.22 g in 0.75 g DMAc and 3.00 g water) solution were prepared in advance. The Nafion solution (0.70 g, stirred at 500 rpm and 90 °C) was then slowly diluted with DMAc (0.75 g), water (12.50 g), and the EMImOH solution. After three days, Nafion molecules formed extended-network micelles, turning the mixture into a weak gel resembling yogurt. These micelles were then slowly mixed with the $EMImBF_4$ solution and stirred for three hours at 500 rpm and 90 °C. Finally, the mixture was cast onto a Petri dish and dried at 90 °C for two days at normal pressure before being subjected to vacuum drying for an additional three days, resulting in an 80-μm-thick H-membrane. For comparison, the C-membrane was also prepared as shown in Fig. 3.2b. The mechanical properties and ionic conductivity of these Nafion membranes were evaluated using tensile testing, electrochemical impedance spectroscopy (EIS), and cyclic voltammetry (CV).

3.2.5 Fabrication and Investigation of Actuators

The process for actuator fabrication is described in Fig. 3.3. First, an electrode mixture consisting of 4.0 ml PP dispersion and 0.2 ml DMSO was prepared. Then, 0.8 ml of this mixture was cast onto an H-membrane taped onto a glass plate with polyimide bands. The membrane was dried at 90 °C for 1.5 h, and the same procedure was applied to the other side of the membrane. The resulting sandwich structure, with a thickness of approximately 100 μm, was subjected to vacuum drying for two days at 70 °C before being cut into rectangular actuators. The specific capacitances of the

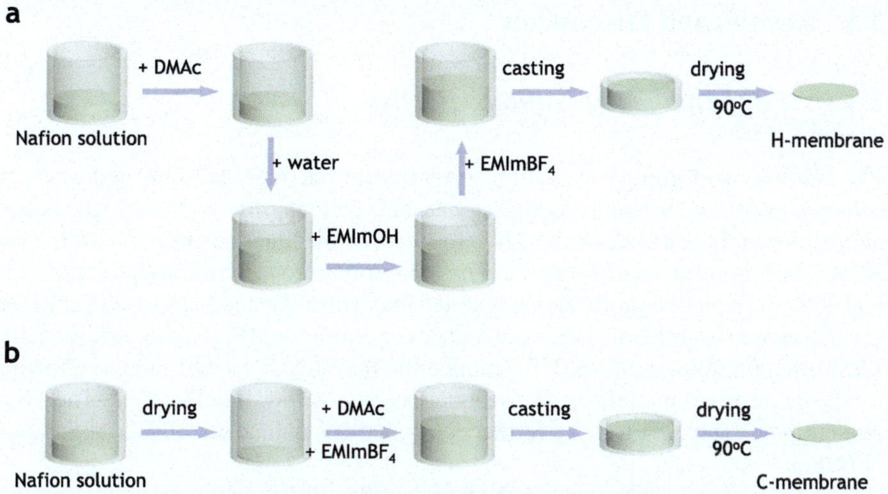

Fig. 3.2 Nafion membrane fabrication [18] from Ref. [18]. Reprinted permission from Springer Nature: **a** new and **b** conventional methods

actuators were measured using cyclic voltammetry (CV). The activated actuators, with two-centimeter free length, were assessed bending deformation using a laser sensor (LK-H080, KEYENCE) and blocking forces using a load cell (LVS-5GA, KYOWA).

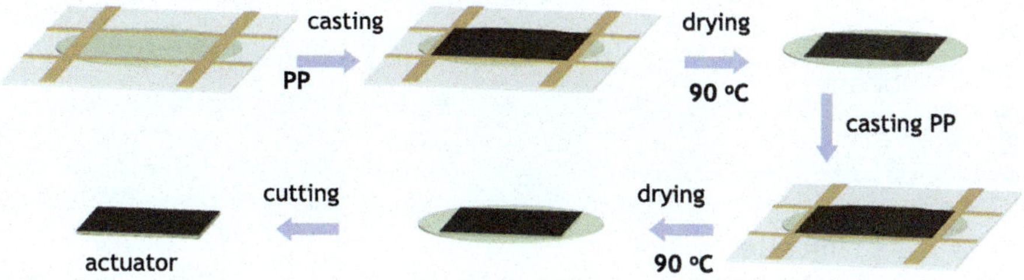

Fig. 3.3 Actuator fabrication [18] from Ref. [18]. Reprinted permission from Springer Nature

3.3 Results and Discussion

3.3.1 TEM Images of Nafion Micelles

The sulfonic acid groups of Nafion were neutralized with caesium hydroxide to enhance visibility during imaging, as caesium cations, due to their high atomic number, strongly scatter electrons [24]. To preserve micellar structures in water, cryo-TEM was employed, as illustrated in Fig. 3.4a. At low concentrations, as depicted in Fig. 3.4b, most Nafion molecules assembled into wormlike micellar networks, which became denser but remained stable at higher concentrations (Fig. 3.4c). Additionally, TEM tomography was utilized to examine the micelles at various angles, allowing for the reconstruction of three-dimensional images, as shown in Fig. 3.4d. The TEM tomography results in Fig. 3.4e further confirmed this unique micellar configuration of Nafion.

The formation of these unique Nafion micelles likely results from intermolecular interactions between Nafion and water molecules. During the diluting steps illustrated in Fig. 3.2a, high temperature, low concentration, and strong repulsion among sulfonate groups promoted the extended coil conformation of Nafion molecules. Over time, these extended coils self-assembled into wormlike micelles due to the hydrophobic effect. Given that long Nafion molecules (with molecular weight ranging from 10^5 to 10^6 Da) have a low density of random hydrophilic side chains (equivalent weight of 1000–1100) [25], assembly occurred simultaneously

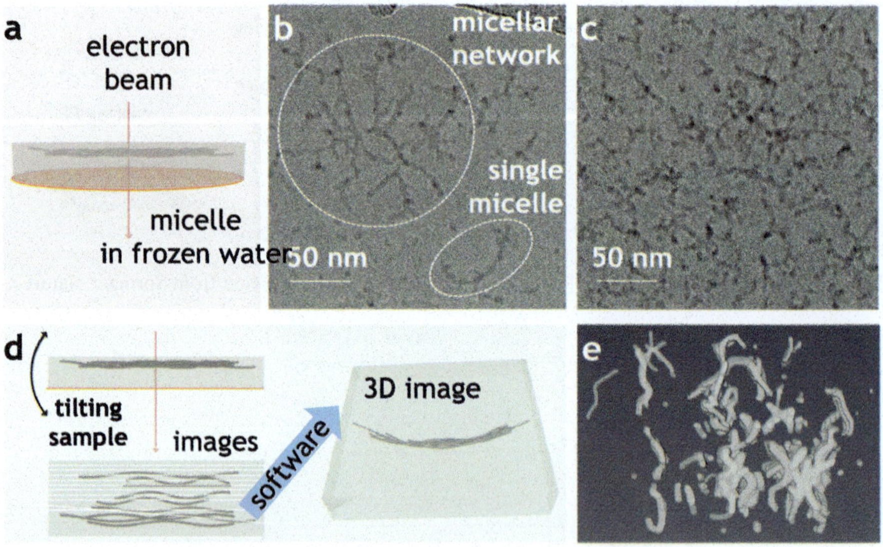

Fig. 3.4 Cryo-TEM images of Nafion micelles [18] from Ref. [18]. Reprinted permission from Springer Nature: **a** Cryo-TEM principle; Nafion micelles in **b** dilute and **c** dense dispersions; **d** cryo-TEM tomography principle and **e** corresponding reconstructed 3D image

among many molecules at different positions, forming extended micellar networks. These micelles remained stable due to the strongly repulsive sulfonate groups on the surfaces. These TEM findings are consistent with the HSP calculations in Table 3.2 and previous reports [14, 16, 24].

3.3.2 Sizes and Surface Charges of Nafion Micelles

The aggregation of Nafion micelles upon the addition of $EMImBF_4$ was investigated by monitoring the particle size using DLS and surface charge via zeta potential measurements. The results are summarized in Fig. 3.5. The addition of $EMImBF_4$

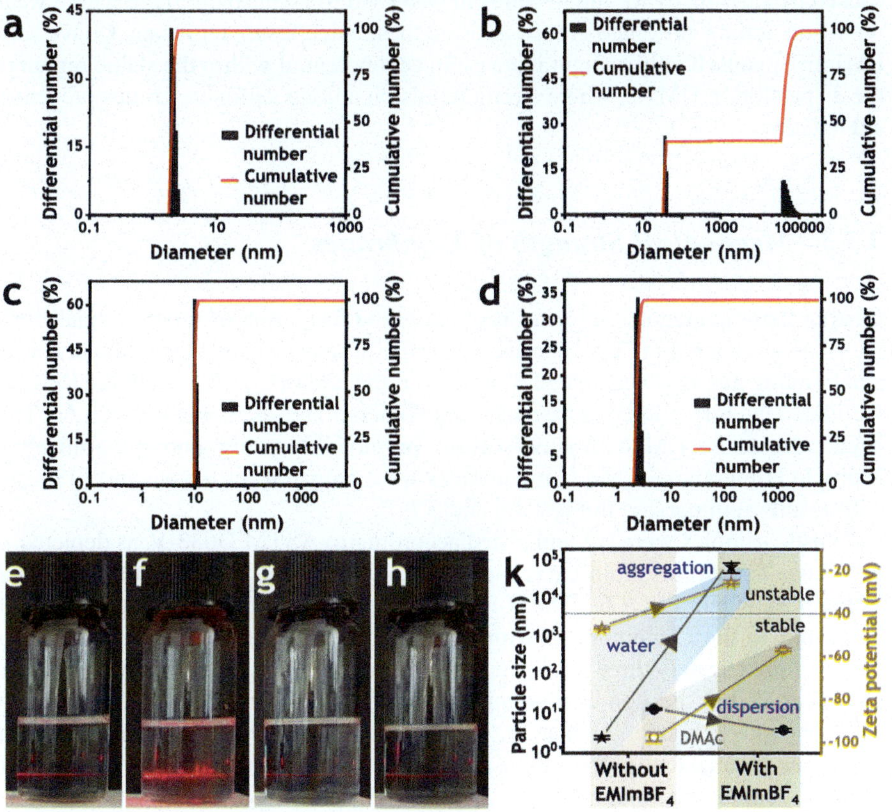

Fig. 3.5 Size and surface charge of Nafion micelles [18] from Ref. [18]. Reprinted permission from Springer Nature: size of Nafion micelles in water **a** before and **b** after adding $EMImBF_4$; size of Nafion dispersions in DMAc **c** before and **d** after adding $EMImBF_4$; optical images of Nafion micelles in water **e** before and **f** after adding $EMImBF_4$; optical images of Nafion dispersions in DMAc **g** before and **h** after adding $EMImBF_4$; **k** particle sizes and zeta potentials

caused the Nafion micelles in water to aggregate, as indicated by the rapid increase in particle sizes showed in the DLS results (Fig. 3.5a and b). However, this ionic liquid had no effect on Nafion dispersion in DMAc, as the particle sizes remained unchanged (Fig. 3.5c and d). Optical images in Fig. 3.5e–h further illustrate these phenomena. Nafion micelles in water without the ionic liquid, as well as Nafion dispersion in DMAc with or without the ionic liquid, scattered less light (Fig. 3.5e, g and h), indicating smaller particle sizes, as described by Rayleigh scattering (Tyndall effect). In contrast, Nafion micelles in water with $EMImBF_4$ exhibited asymmetric and strong light scattering (Fig. 3.5f), indicative of larger particle sizes, as described by Mie scattering.

The zeta potential data in Fig. 3.5k provide an explanation for this aggregation. The negative zeta potentials of all four samples reflect the presence of sulfonate groups on the surfaces of the Nafion particles. The sharp increase in zeta potential caused by $EMImBF_4$ for the micelles in water led to an unstable stage where weak repulsion among sulfonate groups allowed these micelles to aggregate. Conversely, $EMImBF_4$ caused only a slight increase in zeta potential within the stable region of the dispersion in DMAc, where strong repulsion among sulfonate groups prevented aggregation.

3.3.3 Mechanical Strength of Membranes

Casting these aggregated micelles resulted in the formation of separate small fragments, as shown in Fig. 3.6a, because $EMImBF_4$ is a poor solvent that hinders physical bonding among these micelles. However, the addition of 8.5% DMAc to water facilitated the formation of a free-standing H-membrane, as shown in Fig. 3.6b. This occurred because DMAc, a good solvent, promoted bonding among the micelles. Multiple H-membranes and C-membranes were prepared for tensile and EIS tests, with results summarized in Figs. 3.6 and 3.7.

Tensile samples were cut and tested according to ASTM D638-V, as depicted in Fig. 3.6c. The stress–strain curves in Fig. 3.6d were used to calculate mechanical strength, shown in Fig. 3.6e. The H-membrane exhibited lower mechanical strength compared to the C-membrane, with Young's modulus, yield strength, and strain values approximately three times smaller than those of the C-membrane. It is important to note that the non-conducting phase, composed of the tetrafluoroethylene hydrophobic backbones of Nafion, primarily contributes to mechanical strength. The weak mechanical properties of the H-membrane were attributed to the separate non-conducting domains originating from Nafion micelles in water. In contrast, the strong mechanical properties of the C-membrane were due to the continuous non-conducting phase with numerous intramolecular entanglements, resulting from good Nafion dispersion in DMAc. Therefore, these tensile tests further confirmed the structure of Nafion micelles previously demonstrated by cryo-TEM, DLS, and zeta potential measurements.

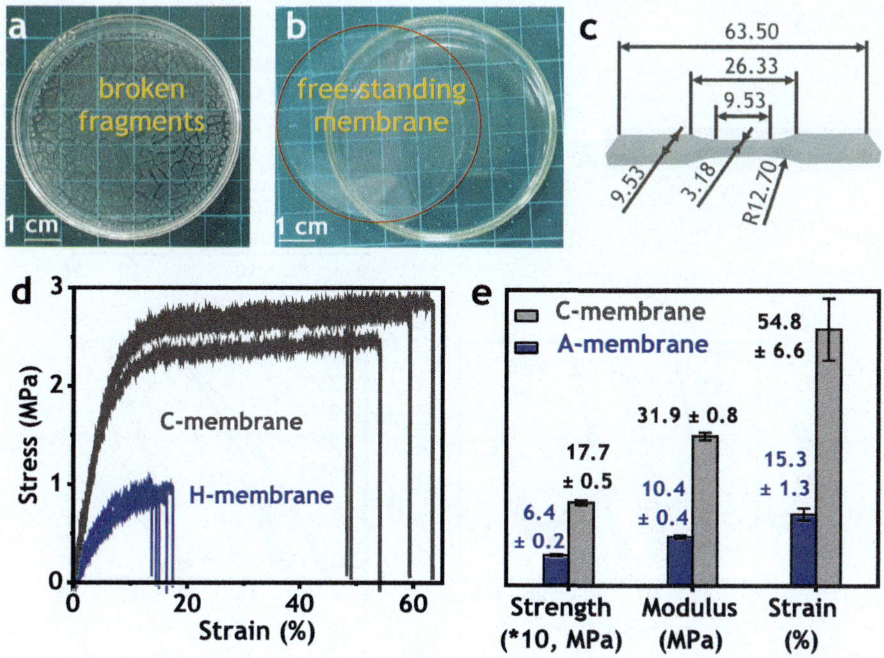

Fig. 3.6 Nafion membrane and its mechanical properties [18] from Ref. [18]. Reprinted permission from Springer Nature: optical images of membranes casting from Nafion micellar dispersions **a** with and **b** without DMAc; **c** dimensions of tensile sample according to ATM D638-V; **d** stress–strain curves; **e** mechanical properties

3.3.4 Ionic Conductivity of Membranes

The membranes were further examined by EIS, with the experimental setup depicted in Fig. 3.7a. The raw EIS data were calibrated by subtracting the short circuit EIS data, as illustrated in Fig. 3.7b. The processed EIS data are plotted in Fig. 3.7c and d, with typical comparisons of EIS spectra of H-membrane and C-membrane shown in Fig. 3.7e. EIS data were used to determine ionic conductivity, as presented in Fig. 3.7f. According to these results, the ionic conductivities of both membranes increased with temperature, following the Arrhenius law. However, across the tested temperatures (24–65 °C), the H-membrane consistently exhibited about three times higher ionic conductivity than the C-membrane. It is noteworthy that the conducting phase, composed of the side chains of Nafion and $EMImBF_4$, primarily contributes to ionic conductivity. Thus, similar to the mechanical strength results discussed earlier, this increase in ionic conductivity was due to the continuous conducting phase of the H-membrane, compared to the separate conducting domains of the C-membrane. These characterizations cross-verified each other, demonstrating the validity of the developed technology for producing the H-membrane with continuous conducting phases

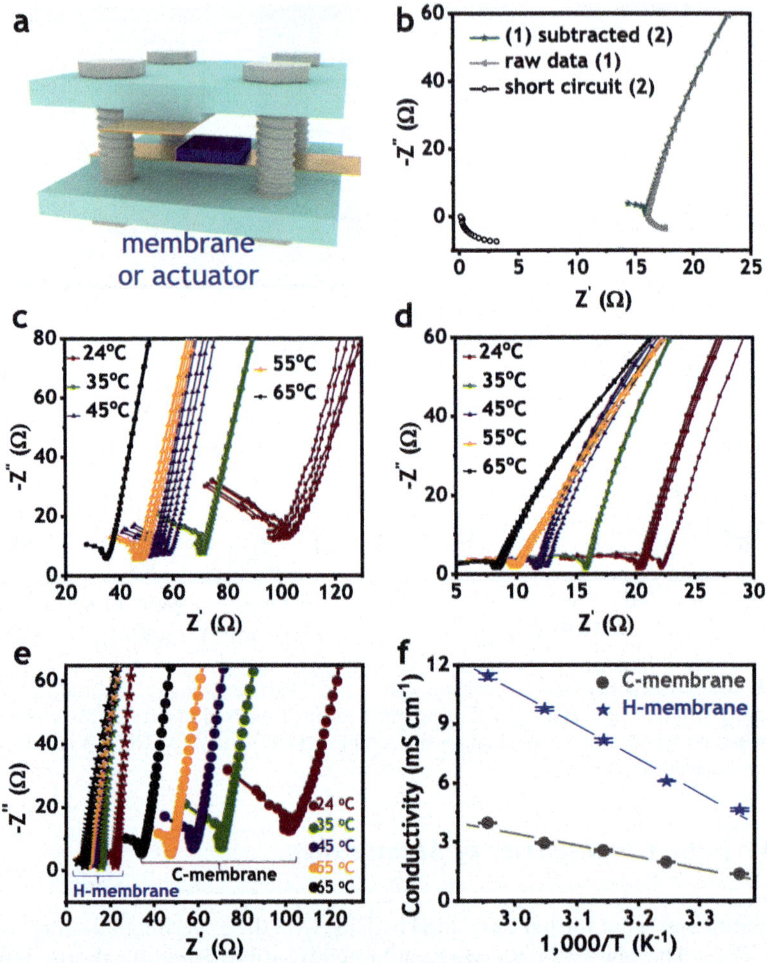

Fig. 3.7 Ionic conductivity [18] from Ref. [18]. Reprinted permission from Springer Nature: **a** EIS measurement setup; **b** calibration of raw EIS data; EIS spectra of **c** C-membrane, **d** H-membrane, and **e** typical comparisons; **f**, ionic conductivity

for high ionic conductivity and adequately connected non-conducting domains for acceptable mechanical strength.

3.3.5 Actuation Performance

Actuators were fabricated from these membranes by casting PP electrodes (Fig. 3.3) and evaluated for specific capacitance using CV tests (Fig. 3.7a). CV curves are shown in Fig. 3.8a–c. As illustrated in Fig. 3.8c, under identical testing conditions and using the same electrodes, the current density of the H-membrane actuator increased more rapidly and to a higher value than that of the C-membrane actuator. This indicates that the former had lower electrical resistivity than the latter. Consequently, this variation in current density further corroborates the difference in ionic conductivity observed in the previous section (Fig. 3.7f), and reinforces the presence of a continuous conducting phase of the H-membrane.

The faster ion supply of the H-membrane compared to the C-membrane during CV tests enabled the former actuator to neutralize more opposite charges in the electrodes within the same time frame. This process was synergistic because the stronger electrical field, due to the higher charge on the electrodes of the H-membrane actuator, in turn, attracted more ions to move faster from the electrolyte to the electrodes. The increase in current density, explained by this synergistic effect, indicates that the H-membrane actuator stored more charge than its C-membrane counterpart, as confirmed by the specific capacitances calculated from the CV data (Fig. 3.8d). Since this synergistic effect is stronger at higher scan rates, increasing frequency from 0.1

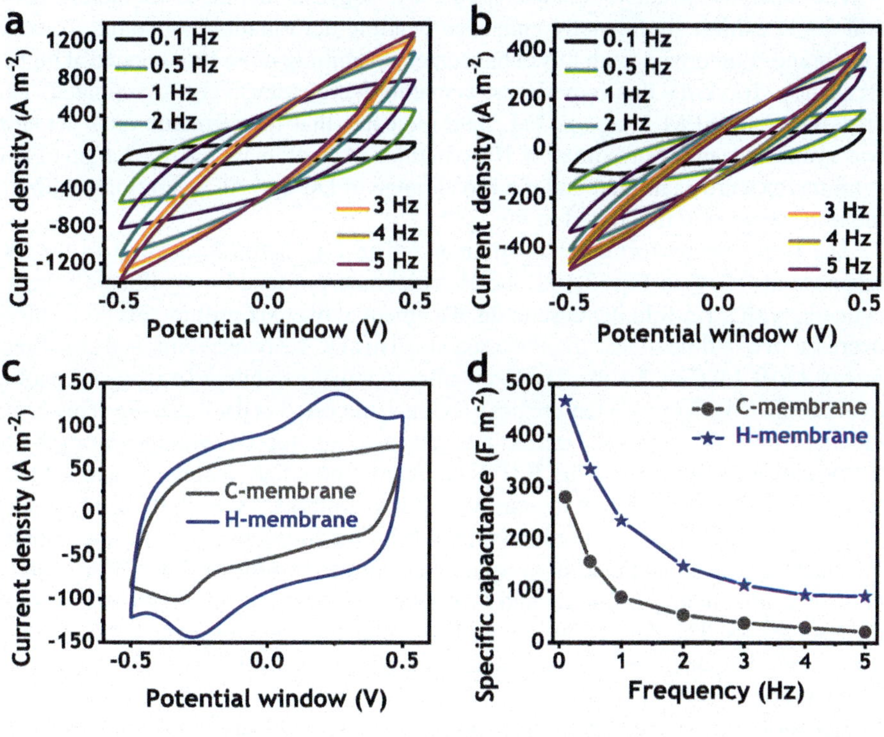

Fig. 3.8 Specific capacitance [18] from Ref. [18]. Reprinted permission from Springer Nature: CV curves of **a** H-membrane and **b** C-membrane; **c** typical comparison of CV curves at 0.2 V/s; **d** specific capacitance

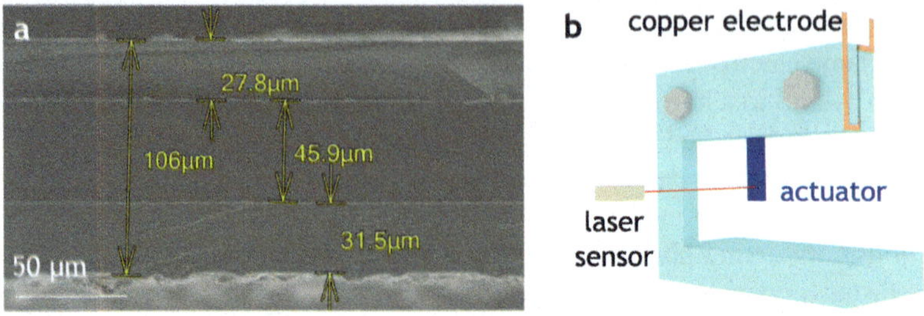

Fig. 3.9 Actuator image and testing setup [18] from Ref. [18]. Reprinted permission from Springer Nature: **a** cross-sectional SEM image of H-membrane actuator; **b** measurement setup for bending deformation

to 5.0 Hz resulted in an accelerated multiplication of specific capacitance in the H-membrane, ranging from 1.7 to 4.5 times higher than that of the C-membrane. This capacitive acceleration clearly proves the importance of the continuous conducting phase in the H-membrane for actuators operating at high frequency.

The capacitive acceleration discussed above suggests that the H-membrane actuator could exhibit significantly enhanced bending deformation compared to its C-membrane counterpart, with this enhancement becoming more pronounced at higher frequency. To verify this hypothesis, actuators were extensively investigated. The cross-sectional SEM image in Fig. 3.9a indicates that the H-membranes actuator was symmetric and approximately 100 µm thick. Actuators with a free length of two centimetres were examined under a broad range of DC and AC conditions, with the measurement setup illustrated in Fig. 3.9b.

The actuation performance is summarized in Fig. 3.10. According to the DC response at 0.3 V in Fig. 3.10a, the H-membrane produced a significantly better actuator, with a 1.5-fold increase in displacement to over six millimeters and tenfold decrease in rise time to less than one second. This DC response remained consistent from 0.1 to 0.3 V (Fig. 3.10b). The H-membrane actuator exhibited no back relaxation over 35 min (Fig. 3.10c), a feature not previously reported in the literature. This large, fast bending was maintained across a broad range of step DC voltages from 0.1 to 0.7 V with 10-s intervals (Fig. 3.10d), showing immediate responses and a linear increase in displacement with increasing voltage amplitude. At each step, the actuator quickly reached and maintained maximum bending during the on-state, and rapidly returned to the original position as soon as the voltage was turned off. Unfortunately, voltages higher than 0.7 V generated deformations beyond the measurement range of the laser sensor. This fast, large, and no-back-relaxation bending over a wide voltage range could enhance the accuracy and reliability of this ionic actuator in various applications.

Actuators were tested across a broad range of AC voltages and frequencies. The response at a square voltage of 1 V and 3 Hz clearly demonstrates the superior quality of the H-membrane, with a 35-fold enhancement in actuation performance (Fig. 3.10e). This improvement was consistently observed over a wide frequency range (0.01–20.0 Hz) and voltage range (0.01–1.00 V), as depicted in Fig. 3.10f

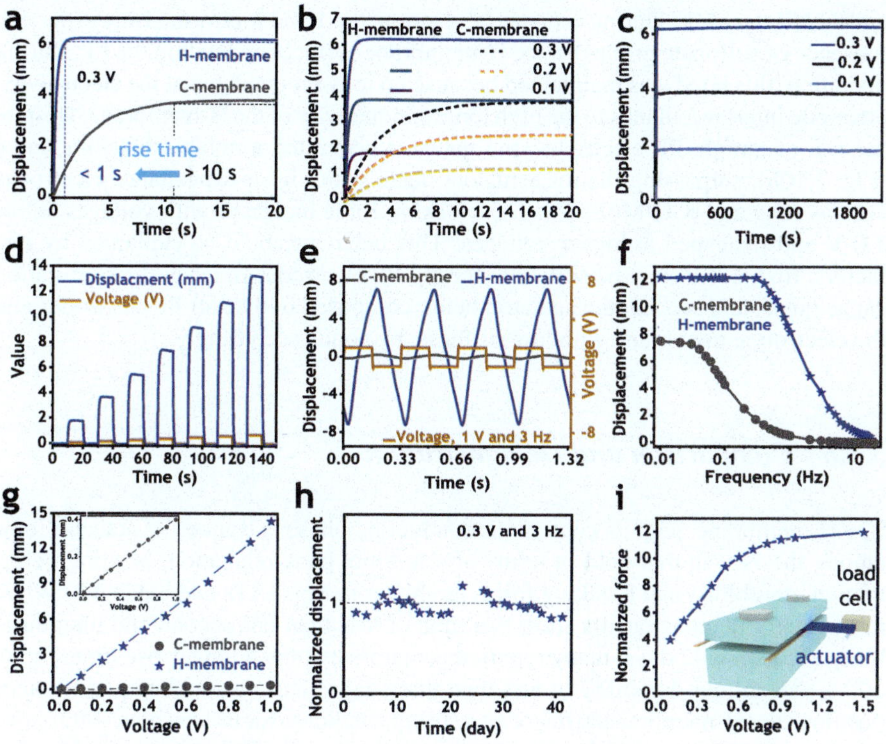

Fig. 3.10 Actuation performance [18] from Ref. [18]. Reprinted permission from Springer Nature: DC responses at **a** 0.3 V, **b** from 0.1 to 0.3 V, **c** over 35 min, and **d** step voltages from 0.1 to 0.7 V; AC response **e** typical bending at 1 V and 3 Hz; bending according to **f** frequency and **g** voltage; **h** durability; **i** blocking force

and g. According to Fig. 3.10f, the H-membrane actuator maintained a maximum displacement of 12.0 mm at frequencies below 0.30 Hz, with a gradual decrease to a noticeable value at 20 Hz. In contrast, its counterpart exhibited a lower maximum bending of 7.5 mm at frequencies below 0.03 Hz, which dropped nearly to zero at frequencies above 2 Hz. As shown in Fig. 3.10g, the H-membrane actuator achieved a sharp bending of up to 15.0 mm at 1 V and 3 Hz, whereas its counterpart exhibited a much smaller deformation of 0.4 mm under the same conditions. This contrast in AC performance is consistent with the difference in DC responses, where the H-membrane actuator showed 1.5 times higher displacement and tenfold shorter rise time as discussed in the previous paragraph. These DC and AC responses collectively confirm the synergistic effect and the capacitive acceleration previously described.

The H-membrane actuator was also evaluated for durability and blocking force. For durability, the actuator was activated at 0.3 V and 3 Hz, with the data normalized to the average over 40 days (Fig. 3.10h). The results indicate exceptional stability over 42 days (equivalent to 11 million cycles), which has not been reported in the literature thus far and could be attributed to the strong bond between the H-membrane and the PP electrode. During actuator fabrication, this bond was formed due to the rich presence of sulfonate groups on the H-membrane surfaces and within the PP. During operation, this bond was maintained because no ions accumulated at the electrolyte–electrode interface, thanks to the high ionic conductivity of the H-membrane. Despite the use of simple PP electrodes, the actuator generated a notable blocking force (Fig. 3.10h), surpassing that of actuators made with higher-capacitance electrodes such as rGO and rGO/MoS_2 [3]. The blocking force increased with voltages below 1.0 V and plateaued at higher voltages. This behavior could be explained by two factors: first, 1.0 V may be sufficient for the PP electrodes to reach their saturation capacitance, and second, the actuator, being composed of the soft PP electrodes and the electrolyte membrane with low stiffness, responds accordingly.

3.3.6 *Comparison and Demonstration*

The H-membrane, with its continuous conducting phase, advanced its actuator to be among the best in this field in terms of rise time, back relaxation, bending deformation, durability, and blocking force, as shown in Fig. 3.11a and Table 3.3. While this actuator demonstrated a short rise time of less than one second and long durability exceeding 42 days, many reported actuators exhibited longer rise times, up to 200 s, and shorter durability of less than three days. This outstanding performance positions the H-membrane actuator as a strong candidate for use in some soft robots. Consequently, the actuator was tested in two demonstration applications. The first involved an inchworm-inspired robot, with its body made from the H-membrane actuator (Fig. 3.11b). This robot was able to crawl over a long distance by continuously bending (Fig. 3.11c). The second demonstration featured a dynamic tensegrity system, in which the H-membrane actuator functioned as a dynamic component that excited other components within the system (Fig. 3.11d). This tensegrity structure was controlled to oscillate over a broad frequency range (Fig. 3.11e). These demonstrations underscore the high quality of the H-membrane and suggest its great potential in future applications.

3.4 Conclusion

This chapter has presented a novel hemostasis-inspired procedure for creating high-performance electrolyte membranes, which significantly enhanced the performance of ionic actuators suitable for use in soft robotics and tensegrity systems. Inspired by the phospholipid assemblies in cell membranes, careful selection of materials and conditions led to the formation of Nafion micelles with ionic surfaces covering non-conducting cores. Mimicking the hemostasis process, these micelles were transformed into electrolyte membranes with a continuous conducting phase for high ionic conductivity and well-connected non-conducting domains for sufficient mechanical

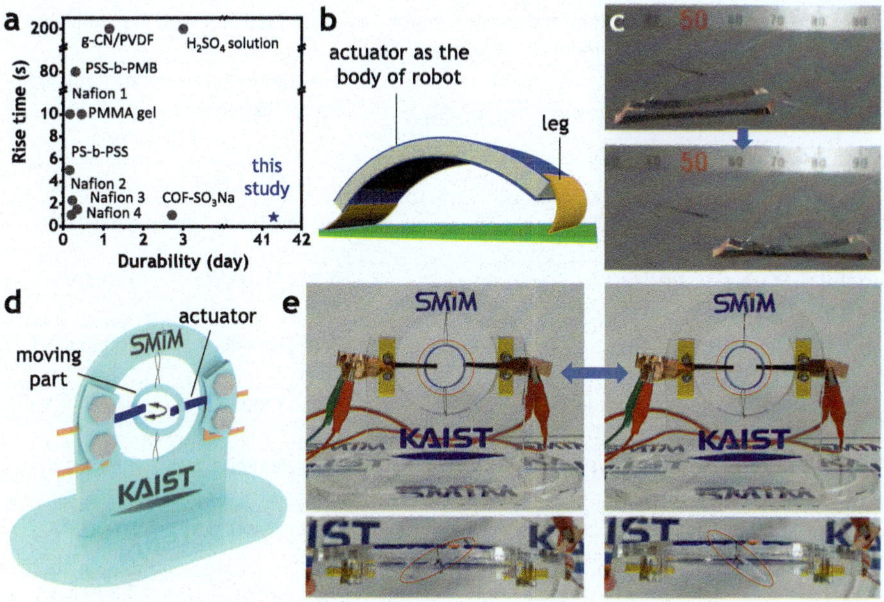

Fig. 3.11 Comparison and demonstrations [18] from Ref. [18]. Reprinted permission from Springer Nature: **a** comparison of rise time and durability, g-CN/PVDF [26], H_2SO_4 solution [27], PSS-b-PMB [9], Nafion 1 [11], PMMA gel [28], PS-b-PSS [5], Nafion 2 [29], Nafion 3 [30], Nafion 4 [12], COF-SO_3Na [31]; inchworm-mimetic robot **b** illustration and **c** optical pictures; dynamic tensegrity system **d** illustration and **c**, optical pictures

strength. This research was systematically monitored through multiple characterizations, from Nafion micelles to the H-membrane and the final actuator. As a result, the actuator's performance was significantly enhanced, achieving a rise time of less than one second, bending of 15 mm at 1 V and 3 Hz, stability over 42 days, and a high blocking force. This performance exceeded that of many reported actuators and was validated through two demonstrations (inchworm-mimetic robots and dynamic tensegrity systems). All data were inter-supported and cross-validated, confirming the effectiveness of the developed technology for producing high-quality H-membranes for use in ionic actuators. This research could advance various fields, including bioinspiration and biomimetics, polymer materials, electrolyte membranes, ionic actuators, and soft robotics. Moreover, the H-membrane and the developed technology presented here could be further applied to related fields such as fuel cells, capacitors, and batteries.

Table 3.3 Comparison of rise time and durability of the developed actuator with the literature [18] from Ref. [18]. Reprinted permission from Springer Nature

Ionic electroactive actuators			Durability [day] (cycle, frequency)	Rise time [s] (voltage)
Electrolyte	Electrode	Thickness [μm]		
H-membrane	PP	100	42 (>10^7, 3 Hz)	0.9 (0.3 V)
0.5M H_2SO_4 solution [27]	MoS_2/Kapton	26	3.0 (8000, 0.03 Hz)	200 (0.6 V)
COF-DT-SO_3Na [31]	PP	60	2.72 (23490, 0.1 Hz)	1 (0.5 V)
g-CN/PVDF/ $EMImBF_4$ [26]	g-CN	85	1.15 (100000, 1 Hz)	200 (3.0 V)
PMMA gel/$LiClO_4$ [28]	Polypyrrole	220	0.46 (120000, 3 Hz)	>10 (1.0 V)
Nafion/$EMImBF_4$ [30]	Ni-MOF/PP	130	0.35 (30000, 1 Hz)	1.5 (1.0 V)
PSS-b-PMB/ $HMImPF_6$ [9]	CNT	70	0.31 (13490, 0.5 Hz)	80 (3.0 V)
PVDF-HFP/ $EMImBF_4$ [32]	PP	140	0.23 (20000, 1 Hz)	2.5 (0.5 V)
Nafion/$EMImBF_4$ [29]	NSCOF/PP	110	0.23 (20000, 1 Hz)	2.3 (0.5 V)
Nafion/$EMImBF_4$ [11]	CTF/PP	80–115	0.17 (15000, 1 Hz)	10 (0.5 V)
PS-b-PSS/ $EMImBF_4$ [5]	Graphene/PP	100	0.16 (14000, 1 Hz)	5 (0.5 V)
Nafion/$EMImBF_4$ [12]	MXene/PP	110	0.21 (18000, 1 Hz)	1 (0.5 V)
	PP	110	0.15 (12600, 1 Hz)	14 (0.5 V)
1M LiCl or NaCl solution [33]	V_2O_5 nanofiber	11	–	>100 (1.0 V)
1M NaCl solution [34]	CNT/PVC/CNT	245	–	>15 (1.0 V)
Other actuators				
Name		Materials		
Fuel-power artificial muscles [35]		NiTi SMA/Pt	–	45
		CNT/Pt/Nafion	–	1800
Polymer composite humid actuator [36]		Polypyrrole	–	300
Photomechanical actuator [37]		CNT/PDMS	–	10
Liquid crystal photoactuator [38]		liquid crystal polymer	–	2400
CNT/thermalplastic actuator [39]		CNT/morthane	–	5
Liquid–crystal elastomer (LCE) thermal actuator [40]		LCE	–	360

(continued)

Table 3.3 (continued)

Ionic electroactive actuators			Durability [day] (cycle, frequency)	Rise time [s] (voltage)
Electrolyte	Electrode	Thickness [μm]		
Liquid–crystal elastomer (LCE) photo actuator [41]		LCE	–	1.3
Dielectric elastomer, applied voltage 4–6 kV [42]		Acrylic, silicone	–	0.001
Shape memory alloys (SMA) actuator [43]		SMA	–	5
CNT yarn muscles [44]		CNT	1.54 (>2 × 10^6, 15 Hz)	–
Fishing line and sewing thread muscles [45]		CNT/nylon	13.89 (1.2 × 10^6, 1 Hz)	–

References

1. Wang J, Gao D, Lee PS (2020) Recent progress in artificial muscles for interactive soft robotics. Adv Mater e2003088. https://doi.org/10.1002/adma.202003088
2. Rich SI, Wood RJ, Majidi C (2018) Untethered soft robotics. Nat Electron 1:102–112. https://doi.org/10.1038/s41928-018-0024-1
3. Manzoor MT et al (2019) Mutually exclusive p-type and n-type hybrid electrode of MoS_2 and graphene for artificial soft touch fingers. Adv Funct Mater. https://doi.org/10.1002/adfm.201905454
4. Shahinpoor M, Bar-Cohen Y, Simpson JO, Smith J (1998) Ionic polymer-metal composites (IPMCs) as biomimetic sensors, actuators and artificial muscles - a review. Smart Mater Struct 7:R15–R30. https://doi.org/10.1088/0964-1726/7/6/001
5. Nguyen VH et al (2019) Electroactive artificial muscles based on functionally antagonistic core-shell polymer electrolyte derived from PS-b-PSS block copolymer. Adv Sci (Weinh) 6:1801196. https://doi.org/10.1002/advs.201801196
6. Tabassian R et al (2019) Graphene mesh for self-sensing ionic soft actuator inspired from mechanoreceptors in human body. Adv Sci. https://doi.org/10.1002/advs.201901711
7. Jo C, Pugal D, Oh IK, Kim KJ, Asaka K (2013) Recent advances in ionic polymer-metal composite actuators and their modeling and applications. Prog Polym Sci 38:1037–1066. https://doi.org/10.1016/j.progpolymsci.2013.04.003
8. Baughman RH (2005) Materials science. Playing nature's game with artificial muscles. Science 308:63–65. https://doi.org/10.1126/science.1099010
9. Kim O, Shin TJ, Park MJ (2013) Fast low-voltage electroactive actuators using nanostructured polymer electrolytes. Nat Commun 4:2208. https://doi.org/10.1038/ncomms3208
10. Kong L, Chen W (2014) Carbon nanotube and graphene-based bioinspired electrochemical actuators. Adv Mater 26:1025–1043. https://doi.org/10.1002/adma.201303432
11. Mahato M et al (2020) CTF-based soft touch actuator for playing electronic piano. Nat Commun 11:5358. https://doi.org/10.1038/s41467-020-19180-3
12. Umrao S et al (2019) MXene artificial muscles based on ionically cross-linked $Ti_3C_2T_x$ electrode for kinetic soft robotics. Sci Robot 4. https://doi.org/10.1126/scirobotics.aaw7797
13. Ma CH et al (2009) Morphology and properties of Nafion membranes prepared by solution casting. Polymer 50:1764–1777. https://doi.org/10.1016/j.polymer.2009.01.060
14. Tarokh A, Karan K, Ponnurangam S (2020) Atomistic MD study of Nafion dispersions: role of solvent and counterion in the aggregate structure, ionic clustering, and acid dissociation. Macromolecules 53:288–301. https://doi.org/10.1021/acs.macromol.9b01663

15. Mabuchi T, Huang SF, Tokumasu T (2020) Nafion ionomer dispersion in mixtures of 1-propanol and water based on the martini coarse-grained model. J Polym Sci 58:487–499. https://doi.org/10.1002/pol.20190101
16. Welch C et al (2012) Nafion in dilute solvent systems: dispersion or solution? ACS Macro Lett 1:1403–1407. https://doi.org/10.1021/mz3005204
17. Kim YS et al (2015) Origin of toughness in dispersion-cast Nafion membranes. Macromolecules 48:2161–2172. https://doi.org/10.1021/ma502538k
18. Nguyen VH et al (2024) Functionally antagonistic polyelectrolyte for electro-ionic soft actuator. Nat Commun 15:435. https://doi.org/10.1038/s41467-024-44719-z
19. Semple JW, Italiano JE Jr, Freedman J (2011) Platelets and the immune continuum. Nat Rev Immunol 11:264–274. https://doi.org/10.1038/nri2956
20. Feinstein MB (1978) Calcium in drug action. In: Feinstein MB (ed) Chapter 9, pp 197–239
21. Hansen CM (2007) Hansen solubility parameters. CRC PressTaylor & Francis Group
22. Rynkowska E et al (2016) Effect of the polar-nonpolar liquid mixtures on pervaporative behavior of perfluorinated sulfonic membranes in lithium form. J Membrane Sci 518:313–327. https://doi.org/10.1016/j.memsci.2016.07.008
23. Weerachanchai P, Wong Y, Lim KH, Tan TT, Lee JM (2014) Determination of solubility parameters of ionic liquids and ionic liquid/solvent mixtures from intrinsic viscosity. ChemPhysChem 15:3580–3591. https://doi.org/10.1002/cphc.201402345
24. Rubatat L, Gebel G, Diat O (2004) Fibrillar structure of Nafion: matching Fourier and real space studies of corresponding films and solutions. Macromolecules 37:7772–7783. https://doi.org/10.1021/ma049683j
25. Mauritz KA, Moore RB (2004) State of understanding of nafion. Chem Rev 104:4535–4585. https://doi.org/10.1021/cr0207123
26. Wu G et al (2015) Graphitic carbon nitride nanosheet electrode-based high-performance ionic actuator. Nat Commun 6:7258. https://doi.org/10.1038/ncomms8258
27. Acerce M, Akdogan EK, Chhowalla M (2017) Metallic molybdenum disulfide nanosheet-based electrochemical actuators. Nature 549:370–373. https://doi.org/10.1038/nature23668
28. Madden JD, Cush RA, Kanigan TS, Hunter IW (2000) Fast contracting polypyrrole actuators. Synth Met 113:185–192. https://doi.org/10.1016/S0379-6779(00)00195-8
29. Mahato M et al (2020) Sulfur- and nitrogen-rich porous pi-conjugated COFs as stable electrode materials for electro-ionic soft actuators. Adv Funct Mater 30. https://doi.org/10.1002/adfm.202003863
30. Mahato M et al (2022) A dual-responsive magnetoactive and electro-ionic soft actuator derived from a nickel-based metal-organic framework. Adv Mater 34:e2203613. https://doi.org/10.1002/adma.202203613
31. Yu F et al (2022) Ionic covalent organic framework based electrolyte for fast-response ultra-low voltage electrochemical actuators. Nat Commun 13:390. https://doi.org/10.1038/s41467-022-28023-2
32. Raza U et al (2022) Micro-structured porous electrolytes for highly responsive ionic soft actuators. Sens Actuators B Chem 352. https://doi.org/10.1016/j.snb.2021.131006
33. Gu G et al (2003) V_2O_5 nanofibre sheet actuators. Nat Mater 2:316–319. https://doi.org/10.1038/nmat880
34. Baughman RH et al (1999) Carbon nanotube actuators. Science 284:1340–1344. https://doi.org/10.1126/science.284.5418.1340
35. Ebron VH et al (2006) Fuel-powered artificial muscles. Science 311:1580–1583. https://doi.org/10.1126/science.1120182
36. Ma M, Guo L, Anderson DG, Langer R (2013) Bio-inspired polymer composite actuator and generator driven by water gradients. Science 339:186–189. https://doi.org/10.1126/science.1230262
37. Ahir SV, Terentjev EM (2005) Photomechanical actuation in polymer-nanotube composites. Nat Mater 4:491–495. https://doi.org/10.1038/nmat1391
38. Iamsaard S et al (2014) Conversion of light into macroscopic helical motion. Nat Chem 6:229–235. https://doi.org/10.1038/nchem.1859

39. Koerner H, Price G, Pearce NA, Alexander M, Vaia RA (2004) Remotely actuated polymer nanocomposites–stress-recovery of carbon-nanotube-filled thermoplastic elastomers. Nat Mater 3:115–120. https://doi.org/10.1038/nmat1059
40. Pei Z et al (2014) Mouldable liquid-crystalline elastomer actuators with exchangeable covalent bonds. Nat Mater 13:36–41. https://doi.org/10.1038/nmat3812
41. Camacho-Lopez M, Finkelmann H, Palffy-Muhoray P, Shelley M (2004) Fast liquid-crystal elastomer swims into the dark. Nat Mater 3:307–310. https://doi.org/10.1038/nmat1118
42. Pelrine R, Kornbluh R, Pei Q, Joseph J (2000) High-speed electrically actuated elastomers with strain greater than 100%. Science 287:836–839. https://doi.org/10.1126/science.287.5454.836
43. Oh S et al (2021) Cooling-accelerated nanowire-nitinol hybrid muscle for versatile prosthetic hand and biomimetic retractable claw. Adv Funct Mater. https://doi.org/10.1002/adfm.202111145
44. Lima MD et al (2012) Electrically, chemically, and photonically powered torsional and tensile actuation of hybrid carbon nanotube yarn muscles. Science 338:928–932. https://doi.org/10.1126/science.1226762
45. Haines CS et al (2014) Artificial muscles from fishing line and sewing thread. Science 343:868–872. https://doi.org/10.1126/science.1246906

Chapter 4
Calcium-Metabolism-Inspired Electrode

4.1 Introduction to Electrode Materials for Ionic Artificial Muscles

Ionic actuators are recognized as promising candidates for artificial muscles in use in soft robotics applications [1–15]. To be practical, these artificial muscles must exhibit large and rapid deformation at low operating voltages, high blocking force, no back relaxation, and long-term durability [16–21]. The typical structure of these actuators consists of a sandwich-like configuration, where an electrolyte membrane is coated with two electrodes. The actuation mechanism is driven by ion rearrangement and mass redistribution when an electrical potential is applied. Key properties for the electrodes include high electrical conductivity, significant capacitance, large surface area, and robust stability [16]. However, materials that possess all these properties are rare.

For example, while graphene and CNTs offer excellent conductivity, they exhibit low capacitance, necessitating high input voltages of around 6 V (Fig. 4.1a and b). Conversely, titanium carbine, although possessing high capacitance, suffers from poor conductivity, leading to slow response time of approximately 0.01 Hz (Fig. 4.1c) [17–22]. Simple composites of these materials, for example graphene flakes and graphene meshes, have proven less effective, likely due to poor contact and significant differences in electronic band structures, which result in slow charge transfer (Fig. 4.1d) [23–31]. Consequently, the development of multifunctional materials for ionic actuators remains a significant challenge.

Insights from calcium metabolism may offer a solution. Calcium, one of the most abundant minerals in the body, plays a crucial role in various physiological processes, including muscle contraction, hemostasis, enzyme function, and bone formation [32]. These processes require precise calcium concentrations to function correctly, but they also cause fluctuations in calcium levels due to continuous consumption and release. Nature resolves this issue through a finely tuned mechanism where calcium is rapidly stabilized by reversible transitions between its hydroxyapatite nanocrystals

V. H. Nguyen, *Bioinspired Multifunctional Nanomaterials for Ionic Artificial Muscles*, Springer Theses, https://doi.org/10.1007/978-3-031-78813-0_4

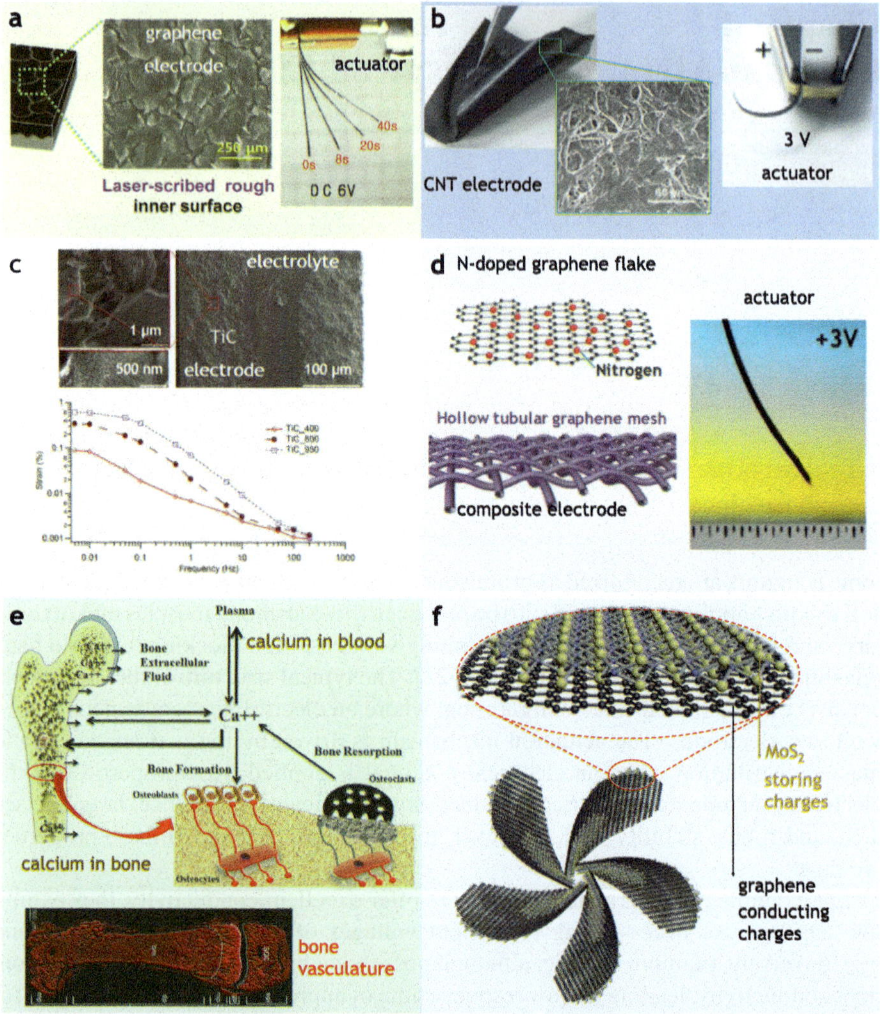

Fig. 4.1 Idea development: some previous electrodes **a** graphene [17] from Ref. [17]. Reprinted with permission from ACS Publications; **b** CNT [35] from Ref. [35]. Reprinted with permission from John Wiley and Sons; **c** TiC [22] from Ref. [22]. Reprinted with permission from Elsevier; **d** graphene mesh/rGO [24] from Ref. [24]. Reprinted with permission from John Wiley and Sons; **e** reversible calcium storage and consumption in the body [33, 34] from Ref. [33]. Reprinted with permission from the Company of Biologists, from Ref. [34]. Reprinted with permission from Springer Nature; **f** calcium-metabolism-inspired MoS_2-rGO [36] from Ref. [36]. Reprinted with permission from John Wiley and Sons

($Ca_{10}(PO_4)_6(OH)_2$) inactive state in bones and active state in the bloodstream [33, 34]. This process is regulated by parathyroid hormone, maintaining blood calcium levels between 8.6 to 10.3 mg/dL, facilitated by a dense vascular network in bones for efficient calcium transport (Fig. 4.1e). This concept of rapid calcium transfer could inspire the development of materials capable of fast charge transfer between capacitive and conductive components.

This chapter introduces a molybdenum disulfide—reduced graphene oxide (MoS_2-rGO) nanocomposite, inspired by the rapid calcium-switching machanism in the body. MoS_2 was selected for its high specific capacitance [37–44], analogous to bones storing calcium, while graphene was chosen for its superior electrical conductivity [45–47], akin to blood vessels transporting calcium. The stable 2D nanomaterials with high surface areas ensure extensive contact, resembling the dense vascular network in bones (Fig. 4.1f). Furthermore, MoS_2 nanoparticles were directly grown on rGO, optimizing the contact area and tuning their electronic bands for efficient charge transfer [48–53]. This bioinspired MoS_2-rGO nanocomposite significantly enhanced actuation performance, increasing bending by a factor of 6.5 compared to rGO at 0.5 V and 1 Hz. As a result, these actuators demonstrated potential for applications such as soft fingers capable of interacting with delicate smartphone and tablet displays.

4.2 Experiment and Characterization

4.2.1 Materials

Key materials used in this chapter include graphite (code #332461), phosphoric acid (H_3PO_4, 85.0 wt.% in water), sulfuric acid (H_2SO_4, 95.0%), potassium permanganate ($KMnO_4$, 99.0%), hydrogen peroxide (H_2O_2, 30.0% in water), thiourea ($CS(NH_2)_2$, 99.0%), and ammonium molybdate tetrahydrate ($(NH_4)_6Mo_7O_{24} \cdot 4H_2O$, 99.0%), N,N-dimethylformamide (DMF, 99.8%), acetonitrile (99.8%), N,N-dimethylacetamide (DMAc, 99.8%), Nafion (D2021, 20.0 wt.% in mixture of lower aliphatic alcohols and water) from Sigma-Aldrich; PEDOT:PSS (PP, CLEVIOS PH 1000) from Heraeus; and 1-ethyl-3-methylimidazolium tetrafluoroborate ($EMImBF_4$, 98.0%) from Io.li.tec; milli-Q type 1 ultrapure water; and some common solvents. All materials were used as received.

4.2.2 MoS_2-rGO Synthesis and Characterization

The MoS_2-rGO was synthesized using two procedures (Fig. 4.2). The first procedure involved the preparation of graphene oxide (GO) through a modifying Tour's method [54], while the second focused on the hydrothermal synthesis of MoS_2-rGO. In the initial procedure (Fig. 4.2a), 3 g graphite was treated with a mixture of 40 ml H_3PO_4 and 360 ml H_2SO_4, followed by the gradual addition of 18 g $KMnO_4$ and subsequent heating at 50 °C for one day. The product was then slowly poured onto ice water, reacted with 10 ml H_2O_2 to remove the $KMnO_4$ residue, and cleaned with water, yielding 5 mg/ml GO dispersion, which was stored at low temperature in the dark for further use. In the second procedure (Fig. 4.2b), three MoS_2-rGO samples (*A*, *B*, *C*) were prepared by varying the amounts of $CS(NH_2)_2$ and $(NH_4)_6Mo_7O_{24} \cdot 4H_2O$ (Table 4.1). For sample *A*, 0.5 g $CS(NH_2)_2$ was dissolved in 10 ml DI water, and 38 mg $(NH_4)_6Mo_7O_{24} \cdot 4H_2O$ was dispersed in 4 ml DI water. These two solutions were mixed with 6 ml GO dispersion before loading into a Teflon lined autoclave, which was kept at 200 °C for one day. The product was filtered, cleaned, and freeze-dried, forming a MoS_2-rGO black powder. The other two samples used the same amount of GO, but more $CS(NH_2)_2$ and $(NH_4)_6Mo_7O_{24} \cdot 4H_2O$ (1.0 g and 76.7 mg for sample *B*, 1.5 g and 114.0 mg for sample *C*, respectively). For comparison, rGO was also synthesized according to the second procedure without $CS(NH_2)_2$ and $(NH_4)_6Mo_7O_{24} \cdot 4H_2O$.

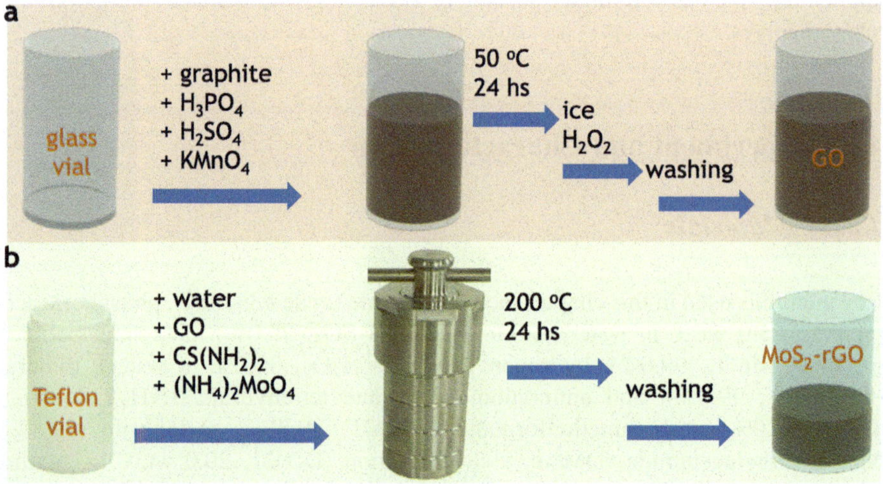

Fig. 4.2 Material synthesis: **a** graphene oxide (GO); **b** MoS_2-rGO

Table 4.1 Precursor ratio for synthesizing MoS_2-rGO [36] from Ref. [36]. Reprinted with permission from John Wiley and Sons

Precursor	MoS_2-rGO		
	A	*B*	*C*
GO [mg]	30	30	30
$CS(NH_2)_2$ [g]	0.5	1.0	1.5
$(NH_4)_6Mo_7O_{24} \cdot 4H_2O$ [g]	38.0	76.7	114.0

The MoS_2-rGO nanocomposite was characterized using field emission scanning electron microscopy (FE-SEM, Magellan400, FEI), field emission transmission electron microscopy (FE-TEM, Tecnai G2 F30 S-Twin, 300 kV, FEI), X-ray photoelectron spectroscopy (XPS, Thermo VG Scientific, K-alpha), high resolution powder X-ray diffractometry (XRD, Rigaku, SmartLab), high resolution Raman analysis (Horiba, LabRAM HR Evolution Visible_NIR), and cyclic voltammetry in 1 M KOH aqueous solution and 0.5 M $EMImBF_4$ acetonitrile solution with a three-electrode system (CV, multichannel potentiostat/galvanostat, Princeton Applied Research).

4.2.3 Fabrication and Investigation of Actuators

Actuators were fabricated through a three-step process (Fig. 4.3). The first step involved preparing the electrode dispersion, where 0.5 ml DMF was added to 10 ml PP and stirred for 30 min before adding 5 mg MoS_2-rGO and stirring for one day (Fig. 4.3a). The second step focused on creating the electrolyte membrane, where 5 g Nafion solution was dried at 80 °C, re-dispersed in 16 ml DMAc at 60 °C for one day, and stirred with 0.6 g $EMImBF_4$ for one day at room temperature before casting 5 ml on a glass petri dish at 85 °C for 6 h (Fig. 4.3b). The third step involved assembling the actuator by applying the electrode mixture to both sides of the electrolyte membrane, followed by drying at 90 °C for 2 h (Fig. 4.3c).

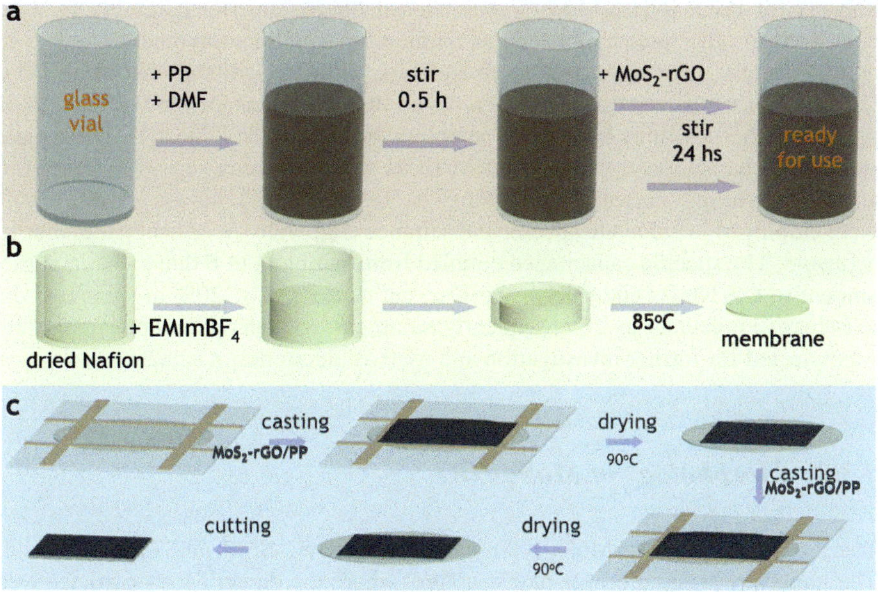

Fig. 4.3 Actuator fabrication: **a** electrode dispersion; **b** electrolyte membrane; **c** actuator. **b**, **c** Nguyen [55] adapted from Ref. [55] with permission from Springer Nature

The MoS_2-rGO/PP electrode's interaction with $EMImBF_4$ was assessed using contact angles measurements (an SEO Phoenix CA system, Surface Electro Optics, Korea). The actuator's performance was characterized through CV with a two-electrode system with the previously mentioned machine, bending deformation measurements with a laser sensor (LK-031, Keyence), and blocking force assessments using a load cell (LVS-5GA, Kyowa). Strains (ε) were calculated from thickness (t), bending curvature (κ), length (l) and bending deformation (δ) of actuators by the following equation.

$$\varepsilon = t\kappa = \frac{2t\delta}{l^2 + \delta^2}$$

4.3 Results and Discussion

4.3.1 Investigation of Precursor Ratio for Synthesizing MoS_2-rGO

The MoS_2-rGO was synthesized using $(NH_4)_6Mo_7O_{24}$, $CS(NH_2)_2$, and GO. $CS(NH_2)_2$ was selected for its thermal decomposition, which releases hydrogen sulfide and ammonia, serving as a sulfur source for MoS_2 and as sulfur-nitrogen dopants for rGO. $(NH_4)_6Mo_7O_{24}$ was chosen for its superior solubility in water compared to other precursors such as ammonium tetrathiomolybdate. The combination of these three precursors in an aqueous solution facilitated the formation of few-layer MoS_2 nanocrystals on rGO. Given that the precursor ratio influences the product properties, three MoS_2-rGO samples were prepared (Table 4.1). These materials were characterized via CV in 1 M KOH solution, using scan rate from 10 to 100 mV/s within a potential window of −0.5 to 0.5 V (Fig. 4.4a–c). These results were employed to calculate specific capacitance (Fig. 4.4d), a critical parameter for actuators. The specific capacitance doubled from sample *A* to *B* due to the increased amount of $CS(NH_2)_2$ and $(NH_4)_6Mo_7O_{24}$, but decreased by 20% in sample *C* due to excessive precursor use. Consequently, sample *B*, with an optimal precursor ratio, was selected for further investigation and used as electrodes in actuators.

4.3.2 Morphology of MoS_2-rGO

The morphology of the MoS_2-rGO was examined via SEM and TEM (Fig. 4.5). The images revealed a flower-like structure, where the darker MoS_2 particles were dispersed regularly across the rGO surface (Fig. 4.5a–c). The MoS_2 particles exhibited a few-layer structure, nanoscale dimensions, and low crystallinity, as indicated

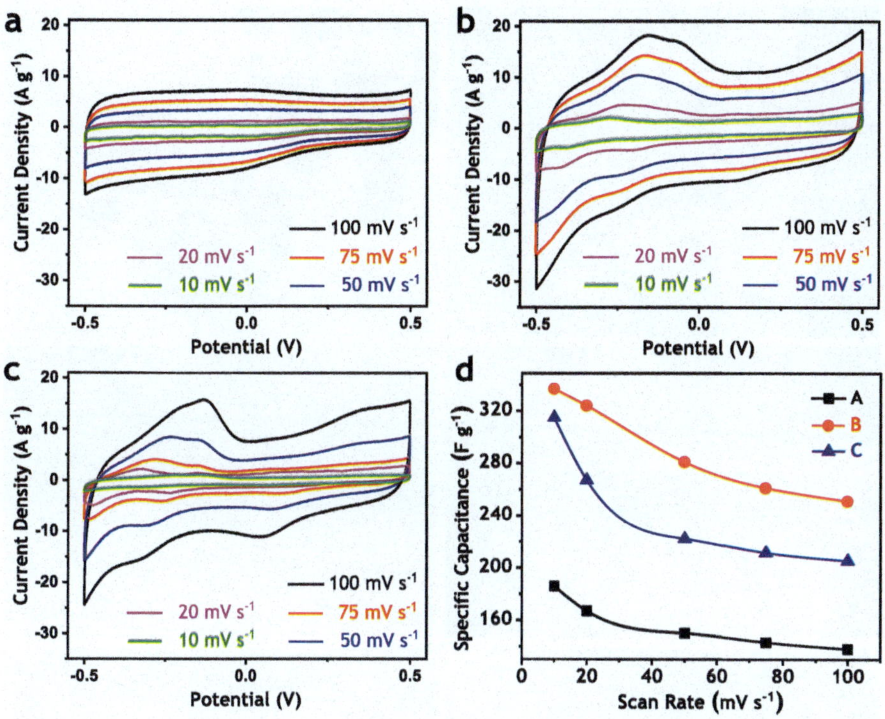

Fig. 4.4 CV curves according to precursor ratios [36] from Ref. [36]. Reprinted with permission from John Wiley and Sons: **a–c** sample *A*, *B*, *C* and **d** their specific capacitance

by the high-resolution TEM and diffraction patterns (Fig. 4.5d). In contrast, rGO appeared as uniform sheets, as shown in TEM images (Fig. 4.5e and f).

This unique MoS_2-rGO structure offers several advantages for actuator applications. First, the flower-like morphology provides a large surface area and open spaces conducive to ion accommodation and interaction. Second, the rGO "petals" facilitate electron transport, which is stored in the MoS_2 nanoparticles situated on their surfaces. Third, the direct contact between MoS_2 and rGO enhances charge transfer between these two components. Finally, the nanoscale size of MoS_2 particles increases their surface area and active site while reducing electrical resistivity, allowing for faster charge and discharge cycles. These features are expected to improve both the speed and magnitude of actuator responses.

4.3.3 Elemental Mapping of MoS_2-rGO

The chemical composition of MoS_2-rGO was analyzed using SEM elemental mapping (Fig. 4.6), which revealed several key characteristics. The material consists

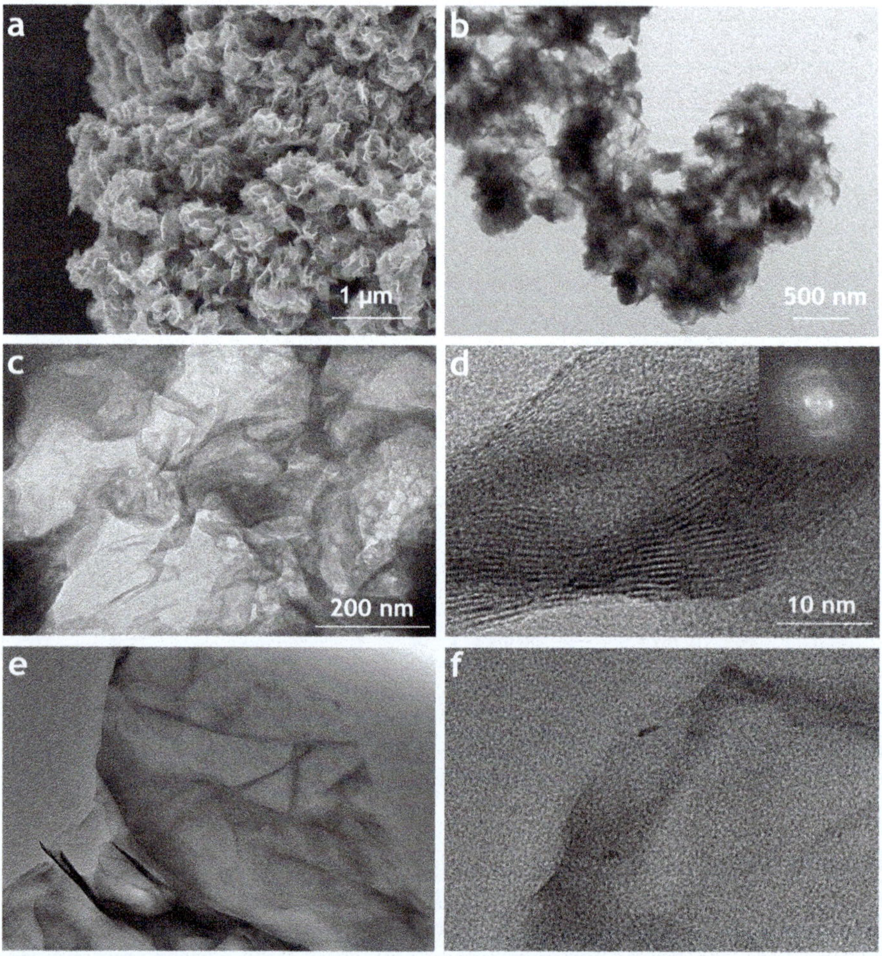

Fig. 4.5 Morphology of MoS_2-rGO [36] from Ref. [36]. Reprinted with permission from John Wiley and Sons: **a** SEM and **b–d** TEM images of MoS_2-rGO; **e**, **f** TEM images of rGO

of five elements: molybdenum, sulfur, oxygen, carbon, and nitrogen. The regular distribution of molybdenum and sulfur indicates uniform scattering of MoS_2 particles (Fig. 4.6b and c). Although carbon is the dominant element, the stronger signals from molybdenum, sulfur, and oxygen suggest that MoS_2 and possible molybdenum oxides are predominantly located on the surface, interacting more intensely with the electron beam than the underlying carbon from the rGO (Fig. 4.6b–e). Additionally, the presence of oxygen is attributed to residue oxygen in the rGO, as the reduction of GO was incomplete under the synthesis conditions. Nitrogen and possible

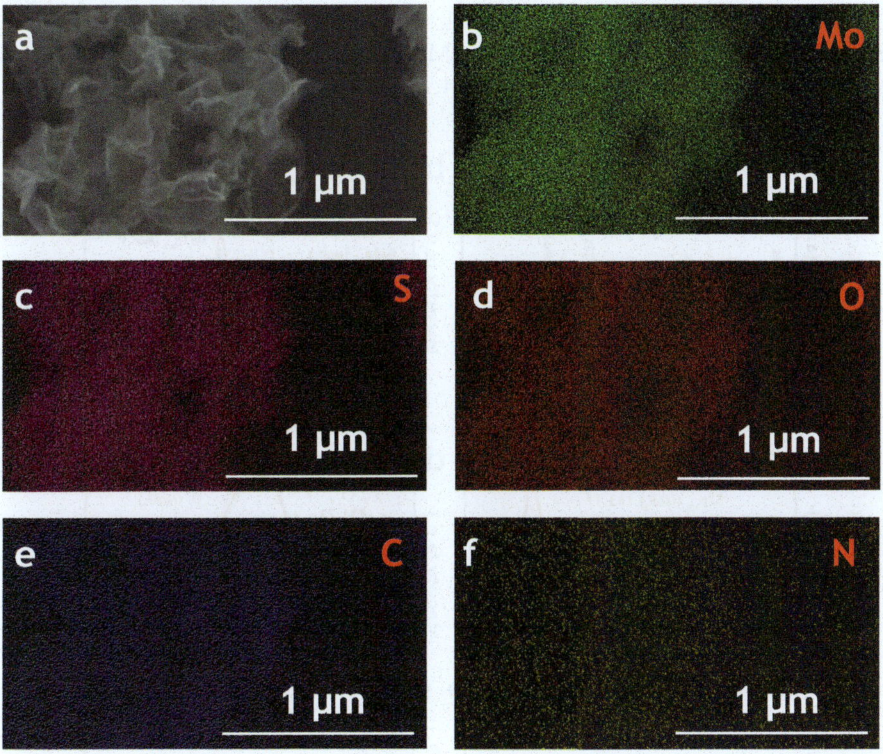

Fig. 4.6 MoS_2-rGO elemental mapping [36] from Ref. [36]. Reprinted with permission from John Wiley and Sons: **a** SEM and corresponding **b** molybdenum, **c** sulfur, **d** carbon, **e** nitrogen, **f** oxygen elemental mapping images

sulfur doping of the rGO is evidenced by their signals (Fig. 4.6c and f), which originated from the decomposition of $CS(NH_2)_2$ in to NH_3 and H_2S. Thus, the MoS_2-rGO consists of thin oxide-doped MoS_2 nanoparticles decorating S,N-codoped rGO sheets.

4.3.4 Chemical Composition of MoS_2-rGO

The elemental composition of MoS_2-rGO was verified through XPS, XRD, and Raman spectroscopy (Fig. 4.7). XPS analysis (Fig. 4.7a) identified the presence of molybdenum, sulfur, carbon, nitrogen, and oxygen [56]. The Mo3d and S2p spectra (Fig. 4.7b and c) exhibited peaks corresponding to MoS_2 (228.23 eV for $Mo3d_{5/2}$, 231.38 eV for $Mo3d_{3/2}$, 161.97 eV for $S2p_{3/2}$, and 163.17 eV for $S2p_{1/2}$) [57] and molybdenum oxides (229.08 eV for $Mo3d_{5/2}$ of MoO_2, 232.38 eV for $Mo3d_{3/2}$ of MoO_2 and $Mo3d_{5/2}$ of MoO_3, and 235.13 eV for $Mo3d_{3/2}$ of MoO_3) [56, 58–60].

These oxides were confirmed by O1s peak splitting at 530.68 eV (lattice oxygen, O_L) and 532.38 eV (surface oxygen, O_S) for MoO_3, and 529.93 eV (O_L) and 531.73 eV (oxygen vacancies, Ov) for MoO_2 (Fig. 4.7d).

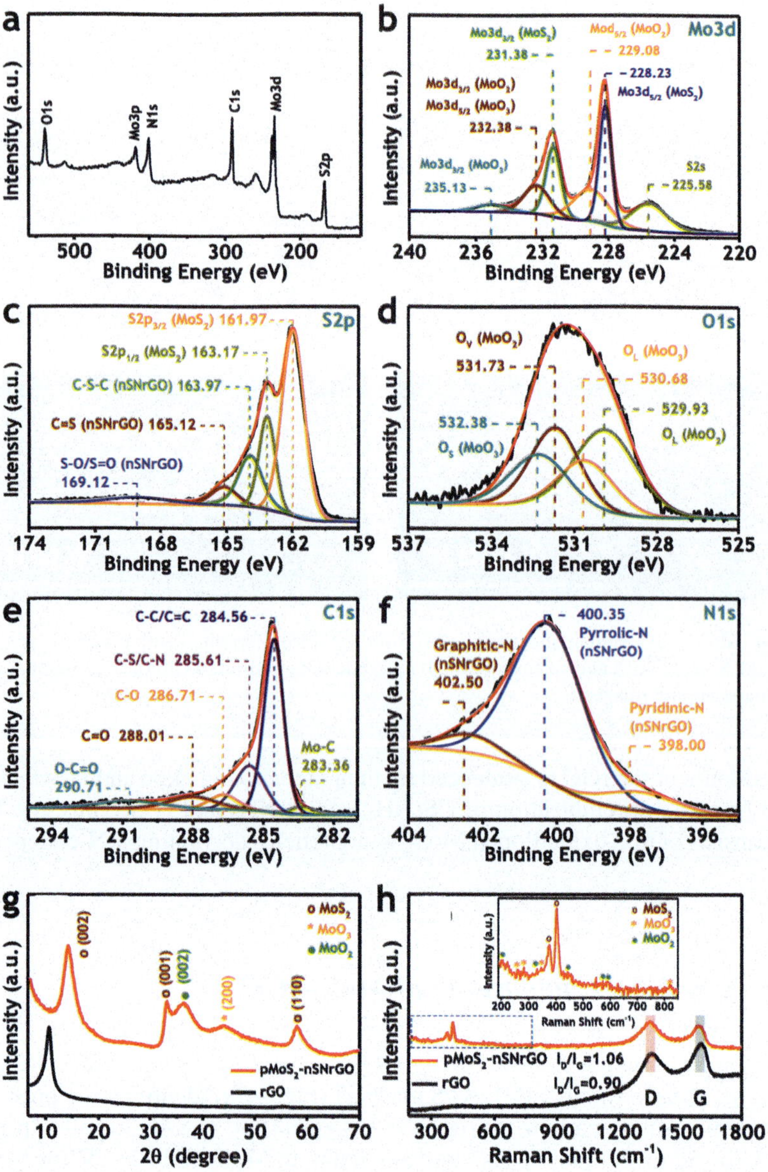

Fig. 4.7 Chemical composition of MoS_2-rGO [36] from Ref. [36]. Reprinted with permission from John Wiley and Sons: **a** survey, **b** Mo3d, **c** S2p, **d** O1s, **e** C1s, and **f** N1s XPS spectra; **g** XRD pattern; **h** Raman spectra

Both XRD and Raman spectroscopy also confirmed the presence of molybdenum oxides (Fig. 4.7g–h). XRD analysis revealed three peaks at 14.2°, 33.2°, and 59° corresponding to the [002], [100], and [110] planes of MoS_2, while Raman spectroscopy detected two vibrations for MoS_2 at 376.6 cm^{-1} $\left(E^1_{2g}\right)$ and 403.3 cm^{-1} $\left(A^1_g\right)$. Despite the dominant MoS_2 peaks, molybdenum oxide peaks were still visible in the XRD spectrum (for example, 37° for the [020] plane of MoO_2; broad shoulder at 45° for the [200] and [002] planes of MoO_3) and in Raman results (207, 346, 456, 572, and 584 cm^{-1} for MoO_2; 248, 283, 352, and 824 cm^{-1} for MoO_3) [56, 59, 61, 62]. The broad shoulders and low intensities of these peaks indicate low crystalline for both MoS_2 and molybdenum oxides, which correlates with the diffraction patterns in Fig. 4.5c. In this MoS_2-MoO_x hybrid, MoO_x doped MoS_2 to behave like a p-type semiconductor [48–51].

XPS, Raman, and XRD results also confirmed the formation of S,N-co-doped rGO. XPS spectra (Fig. 4.7c, e and f) showed three peaks for S2p (C-S-C at 163.97 eV, C = S at 165.12 eV, and S-O/S = O at 169.12 eV), two peaks for C1s (C-C/C = C at 284.56 eV and C-S/C-N at 285.61 eV), and three peaks for N1s (pyridinic-N at 398.00 eV, pyrrolic-N at 400.35 eV, and graphitic-N at 402.50 eV). The C1s spectrum also displayed peaks for C-O, C = O, and O-C = O at 286.71, 288.01, and 290.71 eV, respectively, corresponding to an O1s peak at 532.38 eV for C = O/O-C = O.

The growth of MoS_2 particles on rGO surfaces reduced restacking, as evidenced by the absence of a peak around 26° for the [002] plane of graphene in the XRD pattern (Fig. 4.7g). Stable MoS_2 particles likely contribute to this weak stacking, which could enhance surface area and create open spaces for ion interaction. The Raman spectrum (Fig. 4.7h) shows that the presence of MoS_2 in rGO increased the D/G band ratio from 0.90 to 1.06, indicating numerous rGO defects due to sulfur and nitrogen doping. These defects likely serve as nuclei for MoS_2 nanoparticle growth, act as active sites for ion attraction, increase the charge storage capacity of rGO, and render it an n-type semiconductor that facilitates electron transfer to p-type MoS_2 [29, 31, 52, 53]. This suggests that MoS_2-rGO exhibits high capacitance.

In addition to their direct physical contact, MoS_2 and rGO are chemically bonded, as indicated by the presence of a C1s peak at 283.36 eV corresponding to the Mo-C bond (Fig. 4.7e) [63]. However, this Mo-C bond was not clearly visible in the Mo3d spectrum, likely due to weak Mo-C peaks (232.5 eV for $3d_{3/2}$ and 229.2 eV for $3d_{5/2}$) being overshadowed by strong Mo-O peaks (229.08 and 232.38 eV, Fig. 4.7b). The physical and chemical bonding at the nanoscale caused a shift to lower binding energies in the Mo3d and S2p peaks (231.38 eV for $Mo3d_{3/2}$, 228.28 eV for $Mo3d_{5/2}$, 163.17 eV for $S2p_{1/2}$, and 161.97 eV for $S2p_{3/2}$) compared to pristine MoS_2 (232.67 eV for $Mo3d_{3/2}$, 229.52 eV for $Mo3d_{5/2}$, 163.53 eV for $S2p_{1/2}$, and 162.35 eV for $S2p_{3/2}$) [53, 57, 64] although the band G in the rGO Raman spectrum (Fig. 4.7h) was normal at 1586 cm^{-1}, possibly due to a neutralization between the upshifting of losing electrons to MoS_2 and the downshifting of n-type sulfur and nitrogen doping [31, 52, 53, 57]. The Mo3d and S2p peak shift suggests that that MoS_2 gained electrons from rGO, indicating efficient charge transfer between the two components.

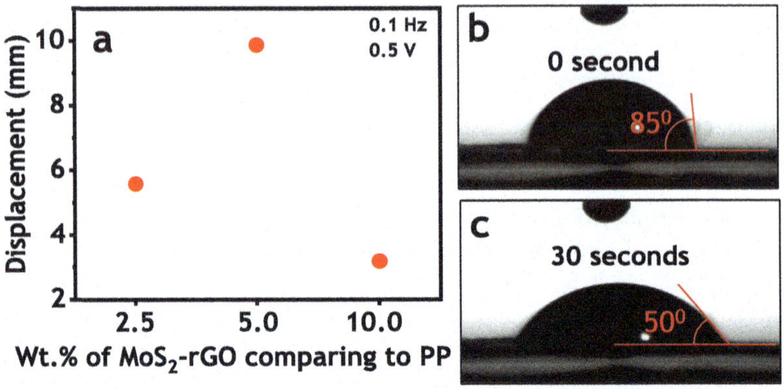

Fig. 4.8 MoS_2-rGO as electrodes of actautors [36] from Ref. [36]. Reprinted with permission from John Wiley and Sons: **a** optimization of MoS_2-rGO/PP for actuator application; **b–c** contact angle of $EMImBF_4$ on the surface of MoS_2-rGO/PP electrode film

4.3.5 MoS_2-rGO/PP Electrode and Its Contact Angle

Since MoS_2-rGO is in powder form and cannot be used on its own, a polymer binder was required to form electrode films for ionic actuators. PP was selected for its good film-forming properties. Additionally, the PEDOT component of PP is a good conductor, enhancing charge transfer, while the PSS component promotes ionic conductivity through its sulfonate groups. In MoS_2-rGO/PP electrodes, too little MoS_2-rGO results in low capacitance, whereas too much can lead to cracks. Initial experiments determined an optimal MoS_2-rGO content of 2.5 to 10 wt.% relative to PP. Within this range, three different ratios were selected for fabricating actuators, whose tip displacements at 0.5 V and 0.1 Hz are summarized in Fig. 4.8a. Increasing the MoS_2-rGO content from 2.5 to 5.0 wt.% approximately doubled the actuator displacement, but further increasing it to 10.0 wt.% reduced displacement by three times, possibly due to poor dispersion of MoS_2-rGO in the electrodes. Therefore, 5.0 wt.% of MoS_2-rGO was an optimal ratio and this electrode was used for further contact angle testes, CV measurements, and actuator fabrication.

The wettability of the MoS_2-rGO/PP electrode with $EMImBF_4$, used in the actuator, was assessed by contact angle measurements (Fig. 4.8b and c). The initial contact angle of about 85° rapidly decreased to 50° within 30 s, indicating strong interaction and adsorption of the ionic liquid by the electrode.

4.3.6 Capacitance of MoS_2-rGO/PP Electrode

CV tests were conducted on MoS_2-rGO/PP and rGO/PP electrodes using a three-electrode system (0.5 M $EMImBF_4$ acetonitrile electrolyte, scan rate from 10 to 100 mV/s, and potential window between −0.5 and 0.5 V), with the results shown in Fig. 4.9a and b. A comparison at 0.1 V/s (Fig. 4.9c) reveals that MoS_2-rGO/

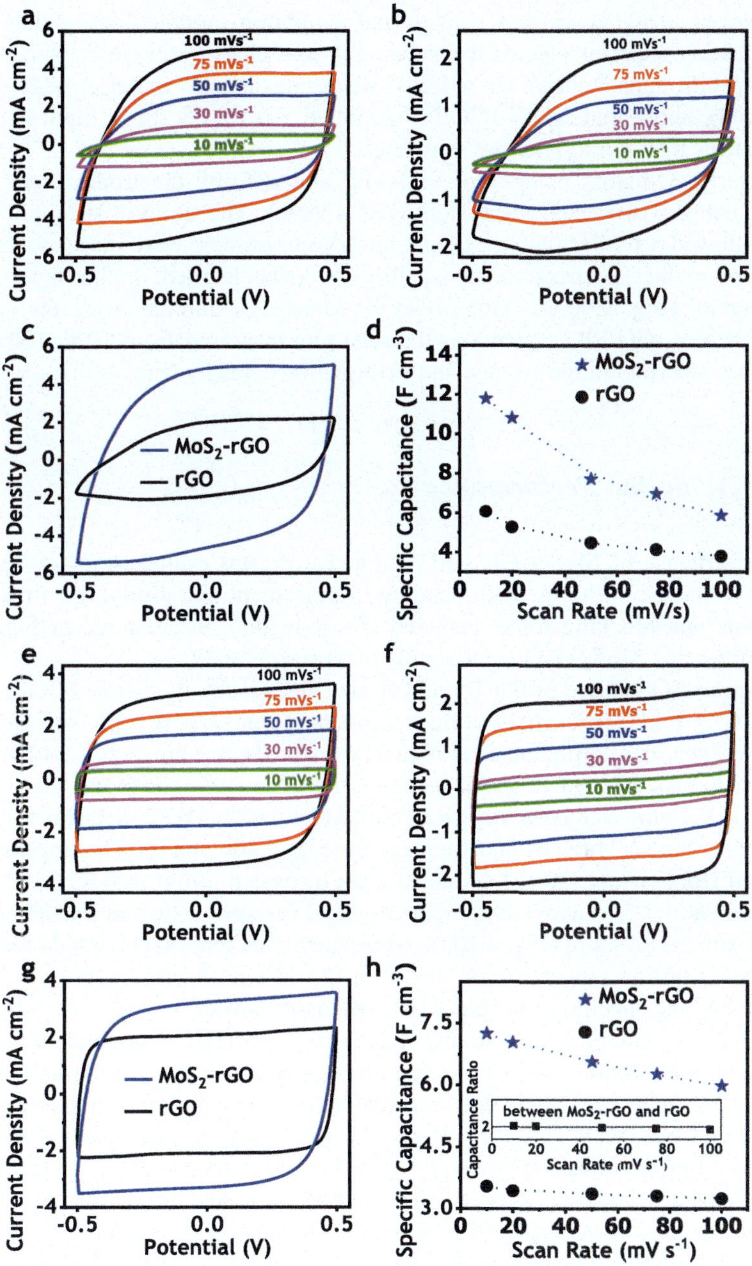

Fig. 4.9 CV curves [36] from Ref. [36]. Reprinted with permission from John Wiley and Sons: in 0.5 M $EMImBF_4/CH_3CN$ **a** MoS_2-rGO/PP, **b** rGO/PP, **c** comparison at 0.1 V/s and **d** specific capacitance; in actuator configuration **e** MoS_2-rGO/PP, **f**, rGO/PP, **g** comparison at 0.1 V/s and **h** specific volumetric capacitance and the inset showing capacitance ratios between two actuators

PP achieved a higher current density and saturation current density than rGO/PP, indicating superior electrical conductivity and charge storage capacity. Volumetric specific capacitances, the ratios between capacitances and electrode volumes (Fig. 4.9d), show that MoS_2-rGO/PP exhibited 2.0 to 1.5 times higher specific capacitances than rGO/PP across scan rates.

Similarly, actuators using MoS_2-rGO/PP and rGO/PP electrodes were tested Vis CV using a two-electrode system (Fig. 4.9e–h). The trend in specific capacitance followed a similar pattern, except the CV curves here were more rectangular, suggesting greater electrochemical stability and ideal electrical double-layer capacitor behavior (Fig. 4.9e–g). MoS_2-rGO/PP actuators exhibited twice the specific capacitance of rGO/PP actuators as the scan rate increased from 10 to 100 mV/s, indicating superior charge storage and deformation capacity (Fig. 4.9h).

4.3.7 Actuation Performance

The performance of MoS_2-rGO and rGO actuators was evaluated under both AC and DC voltage conditions, with bending displacement, durability, rise time, back relaxation, and blocking force analyzed. These results, summarized in Fig. 4.10, demonstrate that MoS_2-rGO significantly outperforms rGO.

Under an AC voltage of 0.5 V and 0.1 Hz (Fig. 4.10a), the MoS_2-rGO actuator showed a 4.5-fold increase in bending displacement compared to rGO. This enhancement was even more pronounced at higher frequencies, reaching a 6.5-fold increase at 1 Hz. Across a broad range of testing conditions (0.1–0.5 V and 0.1–15 Hz), MoS_2-rGO exhibited consistently greater deformations than rGO, with the difference widening at higher voltages and frequencies (Fig. 4.10b and c). The improvement increased from twofold at 0.1 V and 0.1 Hz to over 6.5-fold at 0.5 V and 1 Hz. Remarkably, at 1.5 V and 1 Hz, the MoS_2-rGO actuator achieved a displacement of approximately 10 mm (Fig. 4.10d), surpassing many reported electrode materials over a wide testing range (Fig. 4.10e). The increase in displacement corresponds with the CV and specific capacitance data discussed earlier.

When tested under DC voltage (Fig. 4.10f), the MoS_2-rGO actuator demonstrated a rise time of approximately 10 s, outperforming actuators based on graphene [17] and TiC [22] electrodes, likely due to faster charge transfer. Displacement increased linearly with voltage, consistent with the AC responses. The actuator exhibited no back-relaxation at voltages below 0.2 V and only minor relaxation at voltages above 0.3 V. As back relaxation is more influenced by the electrolyte than the electrode materials, the results are typical for Nafion, an electrolyte with discontinuous ionic conducting phases. The use of a better polymer electrolyte could further reduce back relaxation.

Under DC voltage (Fig. 4.10g), the MoS_2-rGO actuator generated blocking forces approximately 1.5 times greater than those of the rGO actuator across a wide voltage range (0.4–2.0 V). This can be attributed to the higher capacitance of the MoS_2-rGO electrode. Both actuators exhibited rapid increases in blocking force between 0.4 and

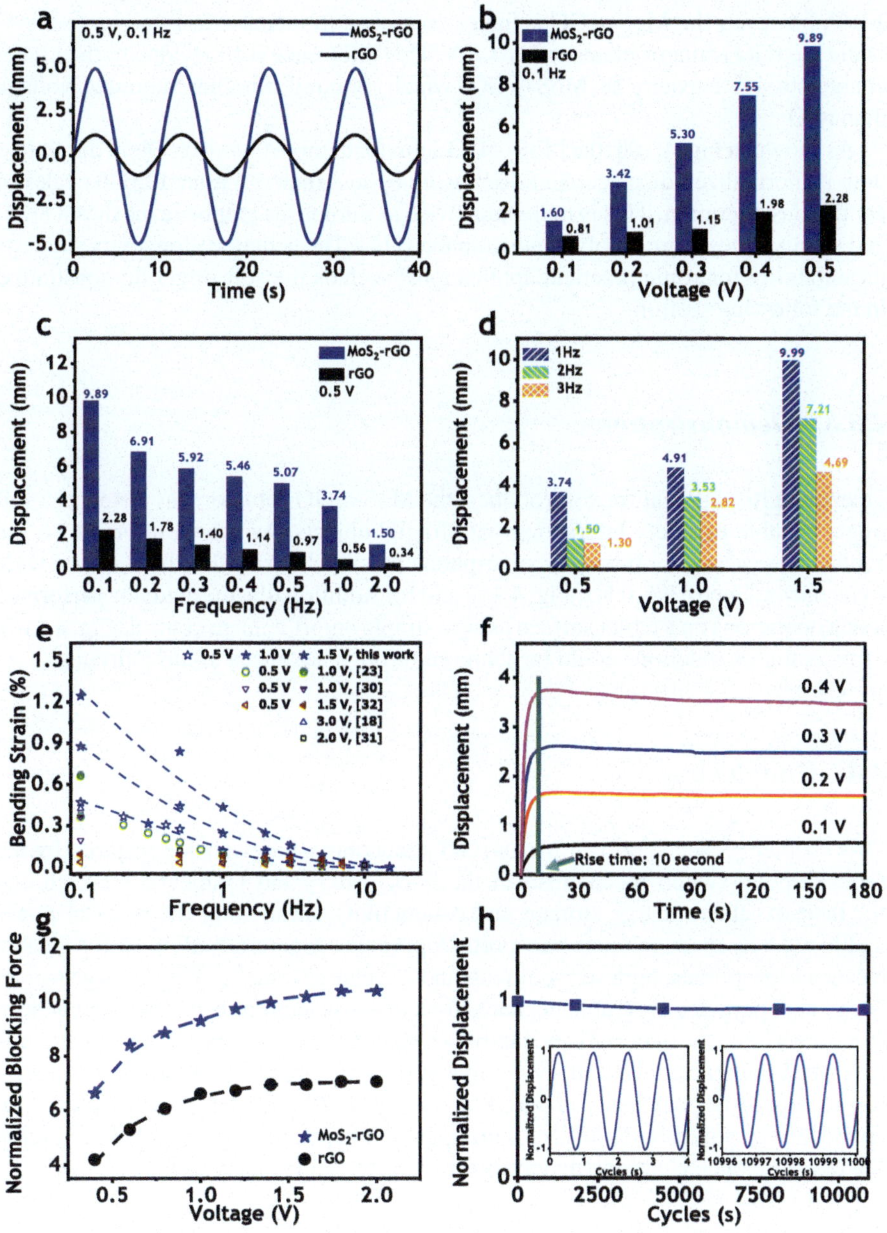

Fig. 4.10 Actuation performance [36] from Ref. [36]. Reprinted with permission from John Wiley and Sons: displacements of actuators **a** at 0.5 V and 0.1 Hz, according to **b** voltage at 0.1 Hz and **c** frequency at 0.5 V; **d** displacements according to high voltages and frequencies; **e** bending strain comparison, rGO [65], rGO [31], Pt [66], SWNTs [21], TiC [22]; **f** DC responses; **g** blocking forces; **h**, durability

1.0 V; however, the MoS_2-rGO actuator continued to increase in force up to 1.8 V, while the rGO actuator plateaued at 1.0 V. This difference is likely due to the higher charge storage capacity of MoS_2-rGO, which required a higher saturated voltage than rGO.

As shown in Fig. 4.10h, the MoS_2-rGO actuator demonstrated excellent durability, with its normalized displacement decreasing by less than 5% after 11,000 cycles of continuous operation. This is more than twice as durable as graphene mesh/rGO [24] and twenty times more durable than graphene [17]. The actuator's longevity suggests its suitability for some practical applications, such as the soft finger demonstration in the following section.

4.3.8 Demonstration

To explore its potential for soft robotics, the MoS_2-rGO actuator was tested as a soft finger capable of safely interacting with fragile objects. As shown in Fig. 4.11, the actuator successfully operated a smartphone by switching its flashlight over a wide frequency range (0.3–8.0 Hz, Fig. 4.11a and b). Additionally, the actuator performed touch operations on a tablet surface using a simple on/off voltage control (Fig. 4.11c). More complex functions could be achieved through advanced control circuits.

4.4 Conclusion

This chapter presents a novel MoS_2-rGO nanocomposite electrode inspired by the rapid calcium-switch mechanism in the body. MoS_2 and Graphene were chosen for their excellent charge storage and conductivity, mimicking the roles of bones and blood vessels in calcium transport. Doping these materials allowed for tunable electronic properties, high surface areas, and fast charge transfer. The direct growth of MoS_2 nanoparticles on graphene, analogous to dense capillaries in bones, enhanced contact between the materials and facilitated charge transfer. This nanocomposite was extensively characterized and shown to significantly improve actuator performance (up to a 6.5-fold increase compared to rGO at 0.5 V and 1 Hz), with successful application in soft robotic fingers. This approach here could be extended to develop high-performance electrode materials for ionic actuators and capacitors.

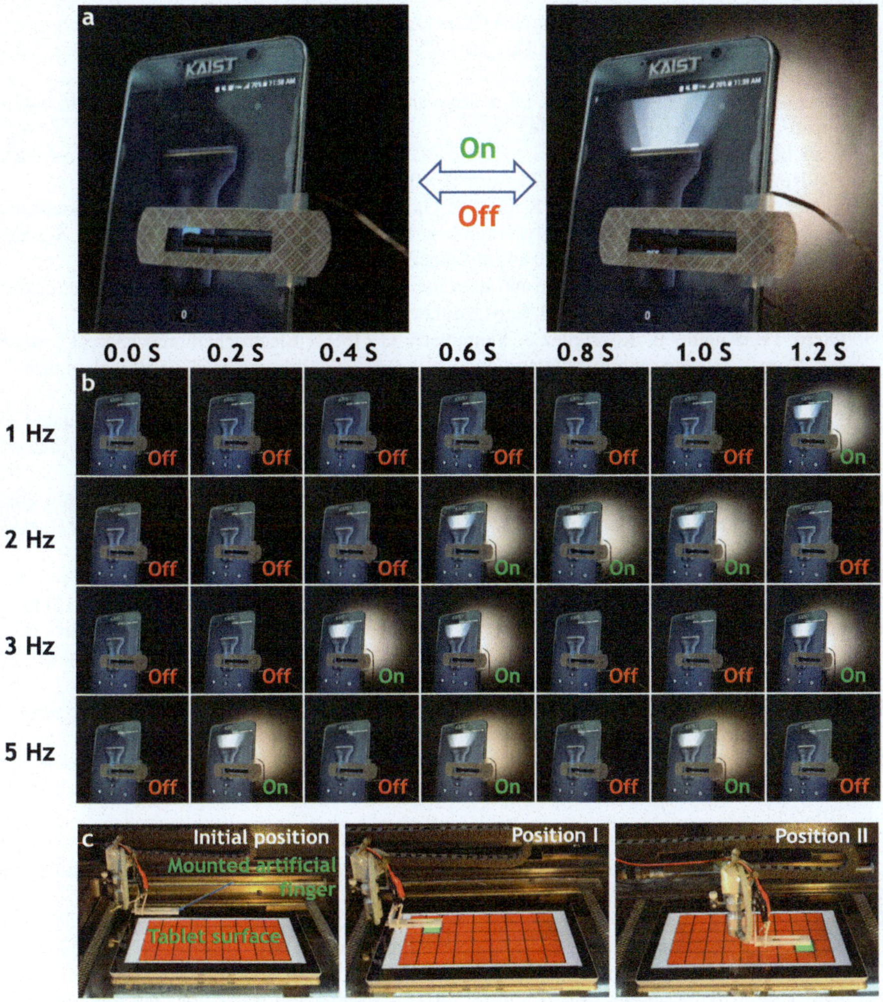

Fig. 4.11 Soft touch finger by MoS_2-rGO actuator [36] from Ref. [36]. Reprinted with permission from John Wiley and Sons: **a** swiching flash light of smartphone; **b** swiching flash light at different frequency; **c** interacting with tablet surface

References

1. Yamada M et al (2008) Photomobile polymer materials: towards light-driven plastic motors. Angew Chem Int Ed Engl 47:4986–4988. https://doi.org/10.1002/anie.200800760
2. Shen Q et al (2016) A multiple-shape memory polymer-metal composite actuator capable of programmable control, creating complex 3D motion of bending, twisting, and oscillation. Sci Rep 6:24462. https://doi.org/10.1038/srep24462

3. Jani JM, Leary M, Subic A, Gibson MA (2014) A review of shape memory alloy research, applications and opportunities. Mater Des 56:1078–1113. https://doi.org/10.1016/j.matdes.2013.11.084
4. Rus D, Tolley MT (2015) Design, fabrication and control of soft robots. Nature 521:467–475. https://doi.org/10.1038/nature14543
5. Cianchetti M, Laschi C, Menciassi A, Dario P (2018) Biomedical applications of soft robotics. Nat Rev Mater 3:143–153. https://doi.org/10.1038/s41578-018-0022-y
6. Zhang QM, Bharti VV, Zhao X (1998) Giant electrostriction and relaxor ferroelectric behavior in electron-irradiated poly(vinylidene fluoride-trifluoroethylene) copolymer. Science 280:2101–2104. https://doi.org/10.1126/science.280.5372.2101
7. Rosset S, Shea HR (2013) Flexible and stretchable electrodes for dielectric elastomer actuators. Appl Phys A-Mater 110:281–307. https://doi.org/10.1007/s00339-012-7402-8
8. Otero TF, Angulo E, Rodríguez J, Santamaría C (1992) Electrochemomechanical properties from a bilayer: polypyrrole/non-conducting and flexible material—artificial muscle. J Electroanal Chem 341:369–375. https://doi.org/10.1016/0022-0728(92)80495-p
9. Aureli M, Kopman V, Porfiri M (2010) Free-locomotion of underwater vehicles actuated by ionic polymer metal composites. IEEE/ASME Trans Mechatron 15:603–614. https://doi.org/10.1109/tmech.2009.2030887
10. Otero TF, Sansiena JM (1998) Soft and wet conducting polymers for artificial muscles. Adv Mater 10:491–494. https://doi.org/10.1002/(SICI)1521-4095(199804)10:6%3c491::AID-ADMA491%3e3.0.CO;2-Q
11. Shahinpoor M, Kim KJ (2005) Ionic polymer–metal composites: IV. Industrial and medical applications. Smart Mater Struct 14:197–214. https://doi.org/10.1088/0964-1726/14/1/020
12. Vidal F, Plesse C, Teyssié D, Chevrot C (2004) Long-life air working conducting semi-IPN/ionic liquid based actuator. Synth Met 142:287–291. https://doi.org/10.1016/j.synthmet.2003.10.005
13. Yeom SW, Oh IK (2009) A biomimetic jellyfish robot based on ionic polymer metal composite actuators. Smart Mater Struct 18. https://doi.org/10.1088/0964-1726/18/8/085002
14. Carpi F, Frediani G, Turco S, De Rossi D (2011) Bioinspired tunable lens with muscle-like electroactive elastomers. Adv Funct Mater 21:4152–4158. https://doi.org/10.1002/adfm.201101253
15. Sotiriou GA, Blattmann CO, Pratsinis SE (2013) Flexible, multifunctional, magnetically actuated nanocomposite films. Adv Funct Mater 23:34–41. https://doi.org/10.1002/adfm.201201371
16. Jo C, Pugal D, Oh I-K, Kim KJ, Asaka K (2013) Recent advances in ionic polymer–metal composite actuators and their modeling and applications. Prog Polym Sci 38:1037–1066. https://doi.org/10.1016/j.progpolymsci.2013.04.003
17. Kim J, Jeon JH, Kim HJ, Lim H, Oh IK (2014) Durable and water-floatable ionic polymer actuator with hydrophobic and asymmetrically laser-scribed reduced graphene oxide paper electrodes. ACS Nano 8:2986–2997. https://doi.org/10.1021/nn500283q
18. Mukai K et al (2009) Highly conductive sheets from millimeter-long single-walled carbon nanotubes and ionic liquids: application to fast-moving, low-voltage electromechanical actuators operable in air. Adv Mater 21:1582-+. https://doi.org/10.1002/adma.200802817
19. Mukai K et al (2008) High performance fully plastic actuator based on ionic-liquid-based bucky gel. Electrochim Acta 53:5555–5562. https://doi.org/10.1016/j.electacta.2008.02.113
20. Kong L, Chen W (2014) Carbon nanotube and graphene-based bioinspired electrochemical actuators. Adv Mater 26:1025–1043. https://doi.org/10.1002/adma.201303432
21. Fukushima T, Asaka K, Kosaka A, Aida T (2005) Fully plastic actuator through layer-by-layer casting with ionic-liquid-based bucky gel. Angew Chem Int Ed Engl 44:2410–2413. https://doi.org/10.1002/anie.200462318
22. Torop J et al (2011) Flexible supercapacitor-like actuator with carbide-derived carbon electrodes. Carbon 49:3113–3119. https://doi.org/10.1016/j.carbon.2011.03.034
23. Tabassian R, Kim J, Nguyen VH, Kotal M, Oh I-K (2018) Actuators: functionally antagonistic hybrid electrode with hollow tubular graphene mesh and nitrogen-doped crumpled graphene

for high-performance ionic soft actuators. Adv Funct Mater 28(5). https://doi.org/10.1002/adfm.201870033
24. Tabassian R, Kim J, Nguyen VH, Kotal M, Oh I-K (2018) Functionally antagonistic hybrid electrode with hollow tubular graphene mesh and nitrogen-doped crumpled graphene for high-performance ionic soft actuators. Adv Funct Mater 28:1705714. https://doi.org/10.1002/adfm.201705714
25. Nguyen VH et al (2018) Electroactive artificial muscles based on functionally antagonistic core-shell polymer electrolyte derived from PS-b-PSS block copolymer. Adv Sci. https://doi.org/10.1002/advs.201801196
26. Kim O, Kim H, Choi UH, Park MJ (2016) One-volt-driven superfast polymer actuators based on single-ion conductors. Nat Commun 7:13576. https://doi.org/10.1038/ncomms13576
27. Kim O, Shin TJ, Park MJ (2013) Fast low-voltage electroactive actuators using nanostructured polymer electrolytes. Nat Commun 4. https://doi.org/10.1038/ncomms3208
28. Roy S, Kim J, Kotal M, Kim KJ, Oh IK (2017) Electroionic antagonistic muscles based on nitrogen-doped carbons derived from poly(triazine-triptycene). Adv Sci (Weinh) 4:1700410. https://doi.org/10.1002/advs.201700410
29. Roy S et al (2019) Collectively exhaustive electrodes based on covalent organic framework and antagonistic co-doping for electroactive ionic artificial muscles. Adv Funct Mater. https://doi.org/10.1002/adfm.201900161
30. Kotal M et al (2018) Highly bendable ionic soft actuator based on nitrogen-enriched 3D hetero-nanostructure electrode. Adv Funct Mater 28. https://doi.org/10.1002/adfm.201802464
31. Kotal M, Kim J, Kim KJ, Oh IK (2016) Sulfur and nitrogen co-doped graphene electrodes for high-performance ionic artificial muscles. Adv Mater 28:1610–1615. https://doi.org/10.1002/adma.201505243
32. Patel VB, Vacek JL, Graves L, Bhattacharya RK (2012) Calcium affects on vascular endpoints. Nutr Metab (Lond) 9:24. https://doi.org/10.1186/1743-7075-9-24
33. Sivaraj KK, Adams RH (2016) Blood vessel formation and function in bone. Development 143:2706–2715. https://doi.org/10.1242/dev.136861
34. Arfat Y, Rani A, Jingping W, Hocart CH (2020) Calcium homeostasis during hibernation and in mechanical environments disrupting calcium homeostasis. J Comp Physiol B 190:1–16. https://doi.org/10.1007/s00360-019-01255-3
35. Mukai K et al (2009) Highly conductive sheets from millimeter-long single-walled carbon nanotubes and ionic liquids: application to fast-moving, low-voltage electromechanical actuators operable in air. Adv Mater 21:1582–1585. https://doi.org/10.1002/adma.200802817
36. Manzoor MT et al (2019) Mutually exclusive p-type and n-type hybrid electrode of MoS_2 and graphene for artificial soft touch fingers. Adv Funct Mater. https://doi.org/10.1002/adfm.201905454
37. Manzoor MT et al (2018) Two-dimensional rGO-MoS_2 hybrid additives for high-performance magnetorheological fluid. Sci Rep 8:12672. https://doi.org/10.1038/s41598-018-30861-4
38. Acerce M, Voiry D, Chhowalla M (2015) Metallic 1T phase MoS_2 nanosheets as supercapacitor electrode materials. Nat Nanotechnol 10:313–318. https://doi.org/10.1038/Nnano.2015.40
39. Soon JM, Loh KP (2007) Electrochemical double-layer capacitance of MoS_2 nanowall films. Electrochem Solid St 10:A250–A254. https://doi.org/10.1149/1.2778851
40. da Silveira Firmiano EG et al (2014) Supercapacitor electrodes obtained by directly bonding 2D MoS_2 on reduced graphene oxide. Adv Energy Mater 4. https://doi.org/10.1002/aenm.201301380
41. Acerce M, Akdogan EK, Chhowalla M (2017) Metallic molybdenum disulfide nanosheet-based electrochemical actuators. Nature 549:370–373. https://doi.org/10.1038/nature23668
42. Zeng Z et al (2011) Single-layer semiconducting nanosheets: high-yield preparation and device fabrication. Angew Chem Int Ed Engl 50:11093–11097. https://doi.org/10.1002/anie.201106004
43. Chhowalla M et al (2013) The chemistry of two-dimensional layered transition metal dichalcogenide nanosheets. Nat Chem 5:263–275. https://doi.org/10.1038/nchem.1589

44. Wang QH, Kalantar-Zadeh K, Kis A, Coleman JN, Strano MS (2012) Electronics and optoelectronics of two-dimensional transition metal dichalcogenides. Nat Nanotechnol 7:699–712. https://doi.org/10.1038/nnano.2012.193
45. Dreyer DR, Park S, Bielawski CW, Ruoff RS (2010) The chemistry of graphene oxide. Chem Soc Rev 39:228–240. https://doi.org/10.1039/b917103g
46. Wang H, Casalongue HS, Liang Y, Dai H (2010) $Ni(OH)_2$ nanoplates grown on graphene as advanced electrochemical pseudocapacitor materials. J Am Chem Soc 132:7472–7477. https://doi.org/10.1021/ja102267j
47. Anasori B, Beidaghi M, Gogotsi Y (2014) Graphene–transition metal oxide hybrid materials. Mater Today 17:253–254. https://doi.org/10.1016/j.mattod.2014.04.043
48. Vattikuti SVP et al (2018) Tiny MoO_3 nanocrystals self-assembled on folded molybdenum disulfide nanosheets via a hydrothermal method for supercapacitor. Mater Res Lett 6:432–441. https://doi.org/10.1080/21663831.2018.1477848
49. Rajeswari J, Kishore PS, Viswanathan B, Varadarajan TK (2009) One-dimensional MoO_2 nanorods for supercapacitor applications. Electrochem Commun 11:572–575. https://doi.org/10.1016/j.elecom.2008.12.050
50. Chuang S et al (2014) MoS_2 P-type transistors and diodes enabled by high work function MoOx contacts. Nano Lett 14:1337–1342. https://doi.org/10.1021/nl4043505
51. Yin Z et al (2014) Preparation of MoS_2-MoO_3 hybrid nanomaterials for light-emitting diodes. Angew Chem Int Ed Engl 53:12560–12565. https://doi.org/10.1002/anie.201402935
52. Yang Z et al (2012) Sulfur-doped graphene as an efficient metal-free cathode catalyst for oxygen reduction. ACS Nano 6:205–211. https://doi.org/10.1021/nn203393d
53. Liu H, Liu Y, Zhu D (2011) Chemical doping of graphene. J Mater Chem 21:3335–3345. https://doi.org/10.1039/c0jm02922j
54. Marcano DC et al (2010) Improved synthesis of graphene oxide. ACS Nano 4:4806–4814. https://doi.org/10.1021/nn1006368
55. Nguyen VH et al (2024) Functionally antagonistic polyelectrolyte for electro-ionic soft actuator. Nat Commun 15:435. https://doi.org/10.1038/s41467-024-44719-z
56. Zhou W et al (2014) MoO_2 nanobelts@nitrogen self-doped MoS_2 nanosheets as effective electrocatalysts for hydrogen evolution reaction. J Mater Chem A 2. https://doi.org/10.1039/c4ta01898b
57. Li G et al Enhanced reversible sodium-ion intercalation by synergistic coupling of few-layered MoS_2 and S-doped graphene. Adv Funct Mater 27. https://doi.org/10.1002/adfm.201702562
58. Jin YS, Shen PK (2015) Nanoflower-like metallic conductive MoO_2 as a high-performance non-precious metal electrocatalyst for the hydrogen evolution reaction. J Mater Chem A 3:20080–20085. https://doi.org/10.1039/c5ta06018d
59. Choi JG, Thompson LT (1996) XPS study of as-prepared and reduced molybdenum oxides. Appl Surf Sci 93:143–149. https://doi.org/10.1016/0169-4332(95)00317-7
60. Baltrusaitis J et al (2015) Generalized molybdenum oxide surface chemical state XPS determination via informed amorphous sample model. Appl Surf Sci 326:151–161. https://doi.org/10.1016/j.apsusc.2014.11.077
61. Bai SL et al (2012) Ultrasonic synthesis of MoO_3 nanorods and their gas sensing properties. Sensor Actuat B-Chem 174:51–58. https://doi.org/10.1016/j.snb.2012.08.015
62. Guha P et al (2019) P-type beta-MoO_2 nanostructures on n-Si by hydrogenation process: synthesis and application towards self-biased UV-visible photodetection. Nanotechnology 30:035204. https://doi.org/10.1088/1361-6528/aaeadc
63. Halim J et al (2016) Synthesis and characterization of 2D molybdenum carbide (MXene). Adv Funct Mater 26:3118–3127. https://doi.org/10.1002/adfm.201505328
64. Zhang ZJ, Zhang J, Xue QJ (1994) Synthesis and characterization of a molybdenum-disulfide nanocluster. J Phys Chem-Us 98:12973–12977. https://doi.org/10.1021/j100100a027

65. Nguyen VH et al (2019) Electroactive artificial muscles based on functionally antagonistic core-shell polymer electrolyte derived from PS-b-PSS block copolymer. Adv Sci (Weinh) 6:1801196. https://doi.org/10.1002/advs.201801196
66. Jung J-H, Jeon J-H, Sridhar V, Oh I-K (2011) Electro-active graphene–Nafion actuators. Carbon 49:1279–1289. https://doi.org/10.1016/j.carbon.2010.11.047

Chapter 5
Bone-Structure-Inspired Electrode

5.1 Introduction to Electrode Materials for Ionic Artificial Muscles

The successful application of ionic actuators in various demonstrations [1–10] underscores their significant potential in soft robotics. However, their current limitations in performance hinder broader practical applications, highlighting the need for further intensive research. Ionic actuators, consisting of two electrodes surrounding an electrolyte membrane, generate bending due to ion currents in the membrane. Therefore, the ideal electrode materials must possess several key properties: high electrical conductivity for fast response, large capacitance for enhanced bending, a substantial surface area for ion interaction, high porosity for efficient ion storage, and stability for long-term durability. Various materials have been explored to improve some of these properties, including noble metals [11–13], conducting polymers [14, 15], CNTs [16, 17], graphene [18, 19], graphitic carbon nitride [20, 21], MXenes [9], metal–organic frameworks (MOFs), and covalent organic frameworks (COFs) [4–6, 10, 22, 23] (Fig. 5.1a–d). However, no single material has yet been identified that possesses all these desirable characteristics.

An intriguing source of inspiration comes from the hierarchical porosity of bones, which feature a unique composite of collagen and hydroxyapatite nanocrystals ($Ca_{10}(PO_4)_6(OH)_2$) within a porous matrix (Fig. 5.1e). Collagen provide micropores (with diameters of 0.6, 1.1, and 1.9 nm) and macropores that house hydroxyapatite, while the interface between these components forms macropores of approximately 4.0 nm. Together, they form the bone matrix, which exhibits a complex porous structure with pore sizes ranging from nanometers to millimeters. This structure supports mechanical functions and creates space for other essential bone components such as blood vessels, living cells, inorganic minerals, and bone marrow [24–28]. This hierarchical porosity serves as a valuable model for designing stable electrodes with similar structured porosity, where the material's walls could conduct and store charges, while the pores provide space for ion storage and interaction. High surface area

V. H. Nguyen, *Bioinspired Multifunctional Nanomaterials for Ionic Artificial Muscles*, Springer Theses, https://doi.org/10.1007/978-3-031-78813-0_5

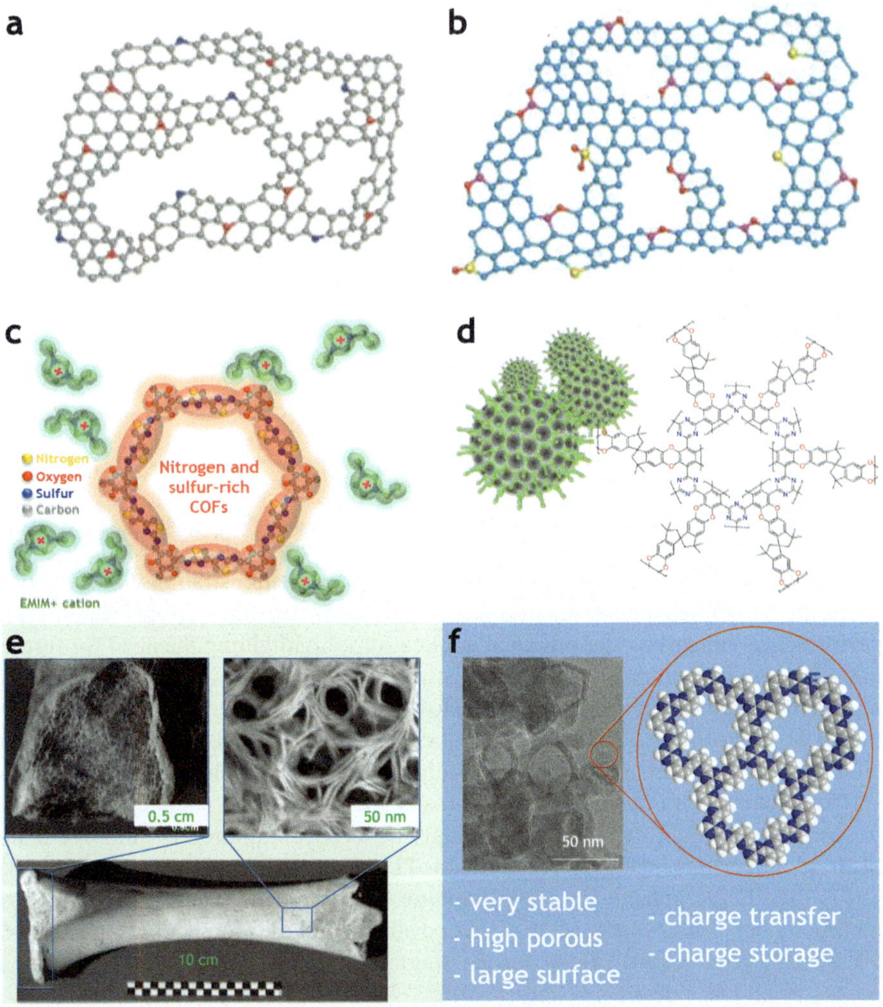

Fig. 5.1 Idea development: COFs used in ionic actuator **a**, PztpOF-derived HPNC [22] from Ref. [22]. Reprinted with permission from John Wiley and Sons; **b**, BS-COF-Cs [23] from Ref. [23]. Reprinted with permission from John Wiley and Sons; **c**, NSCOFs [4] from Ref. [4]. Reprinted with permission from John Wiley and Sons; **d**, PIM-1-derived CTF [6] from Ref. [6]. Reprinted with permission from Springer Nature; **e**, porousity in bones [27, 28] from Ref. [27]. Reprinted with permission from Annual Reviews, from Ref. [28]. Reprinted with permission from Springer Nature; **f**, the new developed CTF [29]

would enhance ion-electrode interaction, and the structural stability would ensure long-term durability.

Porous cross-linked polymers like COFs and MOFs [30, 31] offer a promising avenue for repicating the hierarchical structure of bone. These materials are characterized by ultrahigh porosity, large surface areas [32], structural versatility, and functional tenability [33], making them ideal candidates for electrode materials in ionic actuators. Several electrode systems based on these frameworks have been developed. Functional tunability allows the introduction of heteroatoms to enhance capacitance. For instance, hierarchically porous nitrogen-doped carbon (HPNC) derived from poly(triazine-triptycene) organic framework (PtztpOF) [22] (Fig. 5.1a), boron and sulfur co-doped porous carbon (BS-COF-Cs) derived from a thiophene-based boronate-linked COF [23] (Fig. 5.1b), nitrogen- and sulfur-rich π-conjugated COFs (NSCOFs) [4] (Fig. 5.1c), covalent triazine frameworks (CTFs) derived from polymers of intrinsic microporosity (PIM-1) [6] (Fig. 5.1d), and nickel-based MOFs (Ni-MOFs) [10] have all demonstrated enhanced properties. Additionally, these materials have been combined with other electrode materials to improve electrical conductivity and stability. For example, MOFs have been grown on rGO and MXenes to boost conductivity and durability [34, 35]. These combinations have resulted in actuators with promisting performance. Further improvements in COF and MOF-based electrodes, particularly in CTFs, could be achieved by optimizing chemical structures and pore sizes to incorporate all the necessary properties.

This chapter introduces a novel bone-structure-inspired CTF derived from 2,2′-bipyridine-3,3′-dicarbonitrile (BPDC) (Fig. 5.1f). This monomer was carefully selected to produce a CTF with the five key properties essential for actuator performance. The hierarchically high porosity, large surface area, and robust stability are directly inspired by bone structure and are achieved through the strong covalent bonds within the CTF. The high electrical conductivity is introduced through continuous π-conjugated aromatic rings and partial carbonization. The large capacitance is a result of a high percentage of nitrogen hetoatom, serving as active sites for charge storage and ion interaction. Therefore, this CTF increased actuation bending by over 2.5 times, and the resulting actuator was able to flab the wings of dragonflies in a kinetic art demonstration.

5.2 Experiment and Characterization

5.2.1 Materials

Key materials used in the synthesis of CTF include 2,2′-bipyridine-6,6′-dicarbonitrile (BPDC, 98%) from TCI, anhydrous zinc chloride ($ZnCl_2$, 99.995%), hydrochloric acid (HCl, 37% in water), WHEATON 20 ML ampoule from Sigma-Aldrich. Anhydrous dimethylsulfoxide (DMSO, 99,9%) was from Sigma-Aldrich and poly(3,4-ethylenedioxythiophene)-poly(styrenesulfonic acid) (PP, CLEVIOS PH 1000) was from Heraeus. All materials were employed as received, except for HCl, which was diluted to 1 M solution before use. The electrolyte membranes were prepared according Chap. 3.

5.2.2 CTF Synthesis and Characterization

The synthesis procedure for the BPDC-derived CTF is outlined in Fig. 5.2. BPDC (50 mg, 0.24 mmol) and anhydrous $ZnCl_2$ (165 mg, 1.21 mmol) were added to an ampoule inside a glove box and dried under vacuum at 100 °C for six hours. The ampoule was sealed and heated at a rate of 1 °C per minute to 600 °C, where it was maintained for two days before being cooled naturally to room temperature. The resulting black product was ground and refluxed in water for 12 h to remove $ZnCl_2$, followed by filtering and washing by water. The powder was then stirred with 1 M HCl solution overnight, filtered again, and washed with water until the pH reached 7. After cleaning with acetone to remove residual water, the final product was vacuum-dried at 110 °C overnight, making it ready for further analysis and use as an electrode component in ionic actuators.

The CTF was characterized using solid-state ^{13}C cross-polarization magic angle spinning (CP-MAS) nuclear magnetic resonance (NMR) spectroscopy (Avance Neo 600 spectrometer, Bruker BioSpin), Fourier-transform infrared spectroscopy (FT-IR, with KBr pellets, Nicolet iS50 FT-IR spectrometer, Thermo Fisher Scientific), Raman spectroscopy (excitation laser with a wavelength of 514 nm and a power of 0.5 mW, High-Resolution Raman LabRAM HR Evolution Visible-NIR spectrometer, HORIBA), in-Situ X-ray photoelectron spectroscopy (XPS, at 5×10^{-9} mbar,

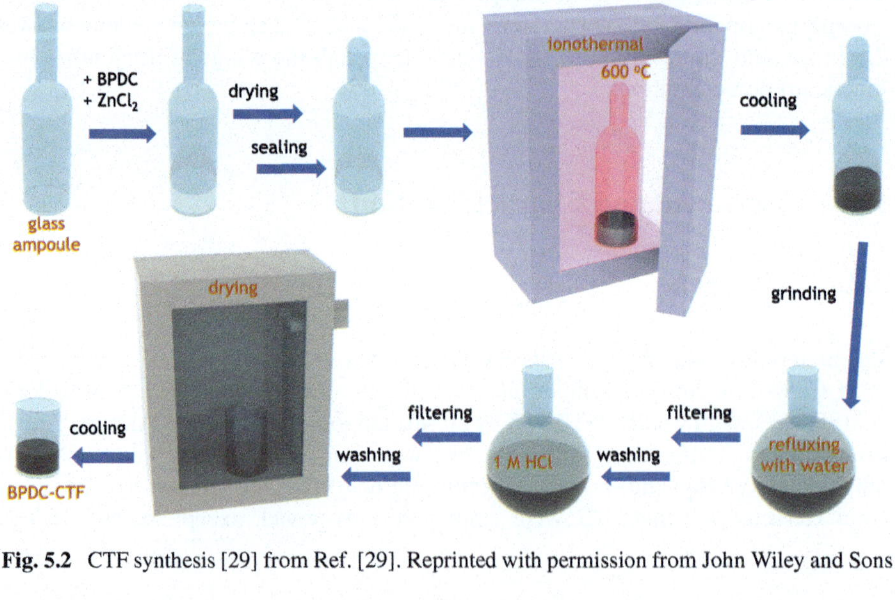

Fig. 5.2 CTF synthesis [29] from Ref. [29]. Reprinted with permission from John Wiley and Sons

Nexsa G2, Thermo Scientific), field-emission scanning electron microscopy (SEM, Magellan400, FEI), field-emission transmission electron microscopy (FE-TEM, Titan cubed G2 60–300, FEI), thermogravimetric analysis (TGA, in N_2 atmosphere, to 1000 °C at 10 °C per minute, TG 209 F1 Libra, NETZSCH TGA), and argon adsorption–desorption isotherms (argon gas, 3Flex. Micromeritics). Specific capacitances were evaluated by cyclic voltammetry (CV, multichannel potentiostat/galvanostat, Princeton Applied Research) at different potential windows and scan rates.

5.2.3 Fabrication and Investigation of Actuators

The preparation of electrode dispersions is illustrated in Fig. 5.3a, the CTF (7.6 mg) was sonicated in DMSO (1.0 ml) for four hours before being gradually added to a PP dispersion (19 ml). This mixture was stirred overnight at 500 rpm and room temperature. A Nafion membrane, approximately 100 μm thick, was prepared following the procedure in Sect. 3.2.4 (Fig. 5.3b). Actuators were fabricated and tested according to the methods described in Sect. 3.2.5, with the CTF/PP composites replacing the PP dispersion (Fig. 5.3c).

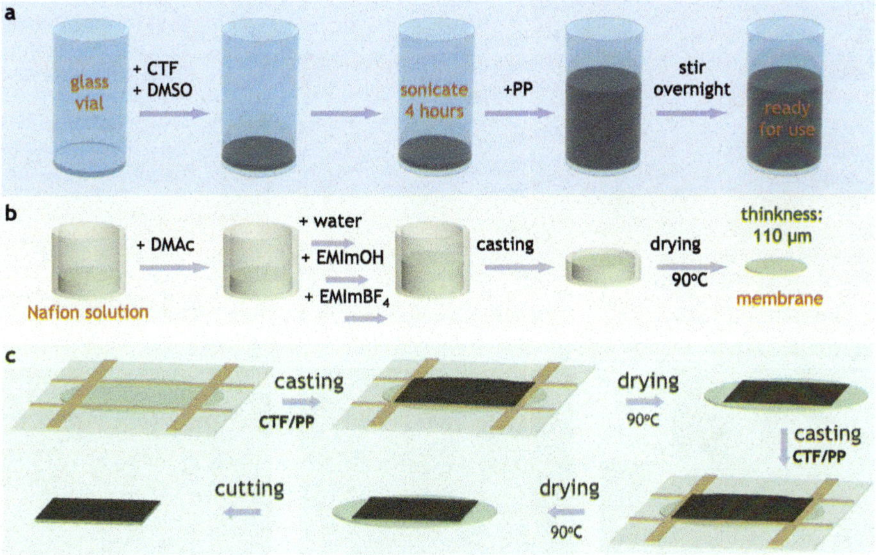

Fig. 5.3 Procedures for: **a** electrode dispersion; **b** electrolyte membrane; **c** actuator fabrication. Figure b, c [36] adapted from Ref. [36] with permission from Springer Nature

5.3 Results and Discussion

5.3.1 TEM Images of CTF

Inspired by hierarchical porosity of bones, this study selected CTFs for their advantageous properties. These include high porosity and large surface areas due to their framework structures, highly chemical and thermal stability from strong aromatic covalent bonds, good charge transfer through π-conjugated aromatic rings, efficient charge storage due to nitrogen-rich heteroatoms, and strong ion interaction from abundant nitrogen-based active sites. The BPDC monomer was specifically chosen for its high nitrogen percentage and molecular configuration, which are conducive to generating high porosity and large surface area. As shown in the inset of Fig. 5.4d, the BPDC monomer, besides having nitrile groups like other CTFs precursors, contains two additional nitrogen atoms in its pyridine rings. The synthesis was conducted at 600 °C, a temperatures that tends to produce higher surface areas [37] and promotes partial carbonization [6], which can enhance electrical conductivity.

The porosity and chemical composition of the CTF were observed by SEM and TEM (Fig. 5.4). The materials exhibited hierarchical porosity, with pore sizes ranging from few nanometers to several hundred nanometers (Fig. 5.4a–c). This structure facilitates ion accommodation and access. Additionally, the presence of amorphous

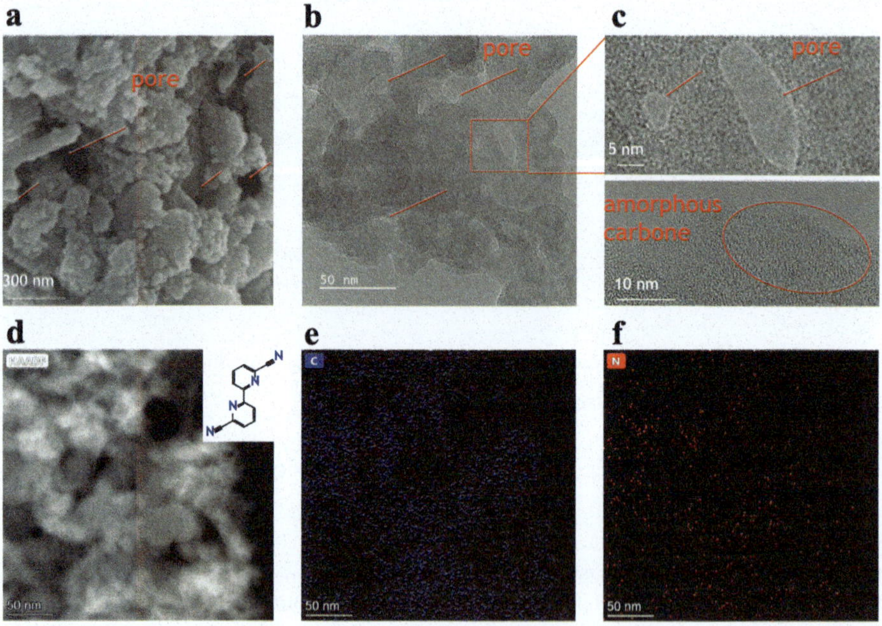

Fig. 5.4 SEM and TEM images [29] from Ref. [29]. Reprinted with permission from John Wiley and Sons: **a** SEM image; **b**–**c** TEM images; **d**–**f** elemental mapping

carbon was observed in the TEM image (Fig. 5.4c), a result of partial carbonization during the synthesis at 600 °C, which is beneficial for improving electrical conductivity. Elemental mapping (Fig. 5.4d–f) confirmed a uniform distribution of carbon and nitrogen across the material, indicating that the nitrogen atoms from the BPDC monomer were retained in the CTF product, forming stable triazine rings and preserving the framework structure.

5.3.2 *Chemical Compositions of CTF*

The formation of triazine rings was confirmed by FT-IR analysis (Fig. 5.5a). Three motable changes indicated successful CTF-synthesis. First, the appearance of a sharp OH peak with broad shoulder suggested the material's polar and porous nature, as it absorbed moisture. Second, the disappearance of the nitrile peak indicated that most nitrile groups were consumed in the reaction. Third, the simplification of aromatic peaks suggested a reduction in molecular configurations due to the formation of triazine rings, which restricted the free rotation of pyridine rings and vice versa.

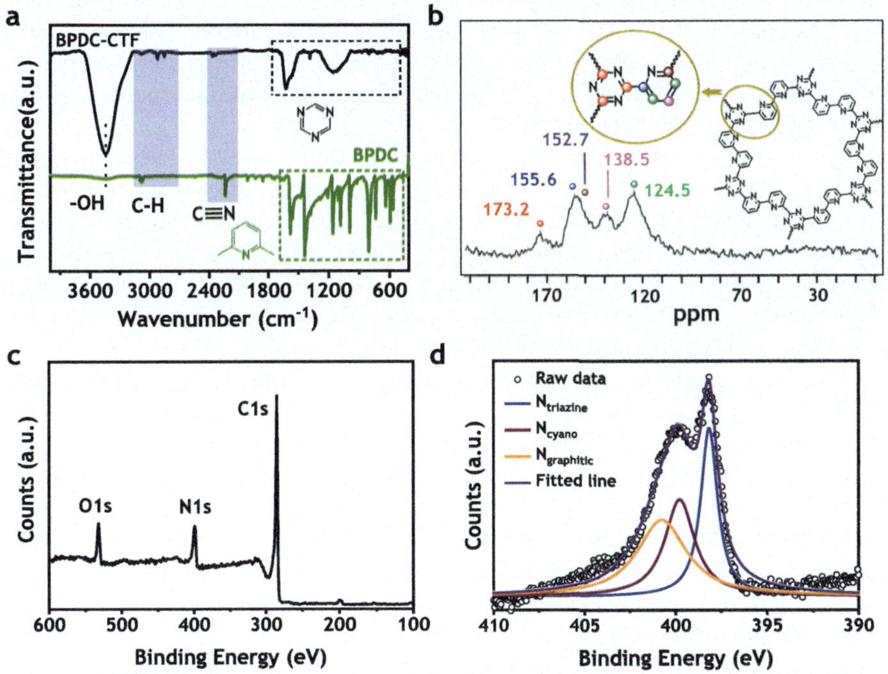

Fig. 5.5 Chemical structure [29] from Ref. [29]. Reprinted with permission from John Wiley and Sons: spectra for **a** FT-IR; **b** solid-state ^{13}C NMR; **c** XPS survey; **d** N1s peak

The chemical structures of pyridine and triazine rings were further validated by ^{13}C NMR and XPS analyses. The NMR spectrum (Fig. 5.5b) displayed peaks consistent with the expected carbon atoms, confirming the presence of pyridine and triazine rings. Furthermore, this NMR result proves that at 600 °C, carbonization happened partially and did not destroy the frameworks and porous structure. This is important because as an electrode material for ionic actuator this CTF should have a proper compromise between porosity and electrical conductivity. The XPS survey (Fig. 5.5c) revealed prominent N1s and C1s peaks, with the presence of O1s peak confirming the absorbed moisture in the previous FT-IR analysis (Fig. 5.5a). The deconvolution of N1s peak (Fig. 5.5d) indicated the presence of triazine ring, cyano (nitrile) group, and graphitic carbon, corroborating the partial carbonization observed in the TEM (Fig. 5.4c).

5.3.3 Surface Area, Stability, and Capacitance of CTF

The specific surface area and the pose size distribution of the CTF were determined through argon adsorption–desorption isotherm analysis at 87 K (Fig. 5.6a). The isotherms displayed Type-I physisorption, indicating a microporous structure, with rapid gas uptake at ultralow pressure suggesting stable microporosity. BET analysis revealed a high specific surface area of 1,800 m^2/g. The pore size distribution (Fig. 5.6b) showed predominant micropores (<2 nm) and some mesopores (2–3 nm), confirming the hierarchical porosity.

The Raman spectrum (Fig. 5.6c) exhibited broad defected D and crystalline G bands at 1350 and 1600 cm^{-1}, respectively, indicative of disordered graphite-like structure in the CTF [38]. The material's stability, attributed to its fully π-conjugated aromatic rings, was further confirmed by TGA data (Fig. 5.6d). The CTF remained stable up to 900 °C, with a minor weight loss observed at 88 °C, corresponding to the evaporation of absorbed moisture, corroborated by the OH peak in the FT-IR (Fig. 5.5a) and the O1s peak in the XPS (Fig. 5.5c).

All experimental characterizations, including SEM, TEM, FT-IR, ^{13}C NMR, XPS, isotherm analysis, Raman, and TGA, demonstrated the successful conversion of the BPDC precursor into a CTF using the ionothermal technique at 600 °C with $ZnCl_2$ as acatalyst. These results revealed that the CTF possessed key properties essential for ionic actuators: hierarchical porosity, a large specific surface area, high stability, partial carbonization, and high nitrogen content. These characteristics suggested the material has high specific capacitance, which was evaluated through CV measurements at different scan rates and potential windows. The CV data (Fig. 5.7a and b) showed nearly rectangular vurves across potential windows of −0.4 to 0.8 V and scan rates ranging from 0.01 to 0.10 V/s, indicating that the CTF behaved similarly to an ideal electrical double-layer capacitor, where no electrochemical reaction occor, and charge neutrality is maintained by ion adsorption at the surface.

However, at voltage beyond −0.5 V, the current density increased rapidly (Fig. 5.7c), likely due to electrochemical reactions involving acetonitrile solvent

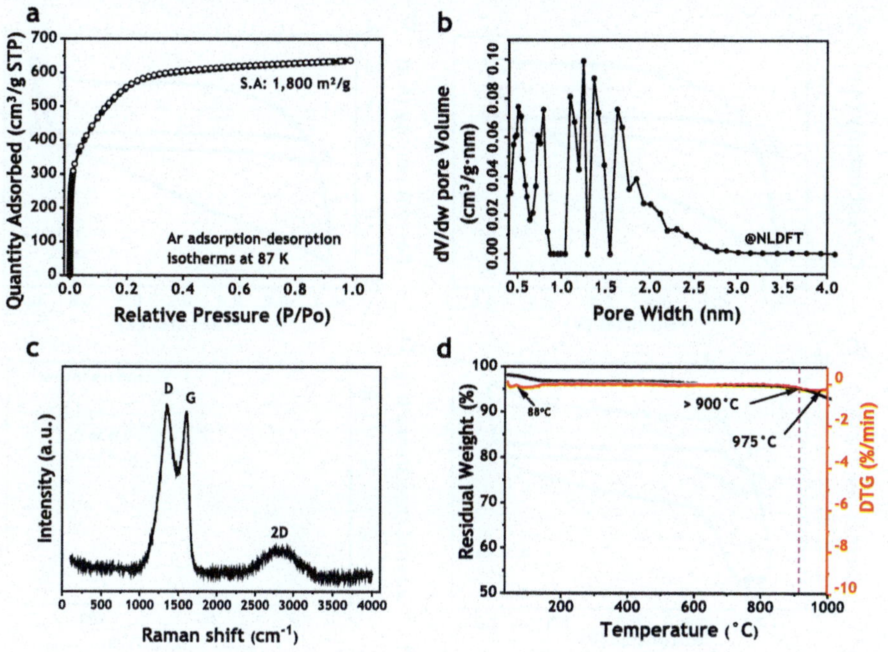

Fig. 5.6 Porousity and stability [29] from Ref. [29]. Reprinted with permission from John Wiley and Sons: **a** argon adsorption–desorption isotherm; **b** pore size distribution; **c** Raman spectrum; **d** TGA

[39], which has a lower stability range compared to pure $EMImBF_4$ (with an intrinsic potential window from −2 to +3.45 V) [39]. Importantly, these results suggest that the CTF and $EMImBF_4$ are suitable for ionic actuators, which typically operate below 2 V and without solvent. Specific capacitance calculation (Fig. 5.7d) reveal a maxinum capacitance of 330 F/g at the scan rate of 0.01 V/s and the potential window from −1 to +1 V, outperforming several reported CTFs and COFs [40, 41]. This high capacitance makes the CTF a promising electrode material for ionic actuators.

The CTF/PP and PP actuators were further investigated by CV using a two-electrode setup with potential window from −0.5 to 0.5 V. According to a typical comparison in Fig. 5.7e, the CTF/PP actuator exhibited 2.2-fold increase in areal capacitance compared to PP, with values of 541 F/m^{-2} versus 247 F/m^{-2}, respectively. This enhancement was consistent across a wide frequency range from 0.01 to 5 Hz (Fig. 5.7f), confirmed the results presenting in Fig. 5.7a–d, and is attributed to the material's high porosity, large surface area, superior electrical conductivity, and nitrogen-rich content.

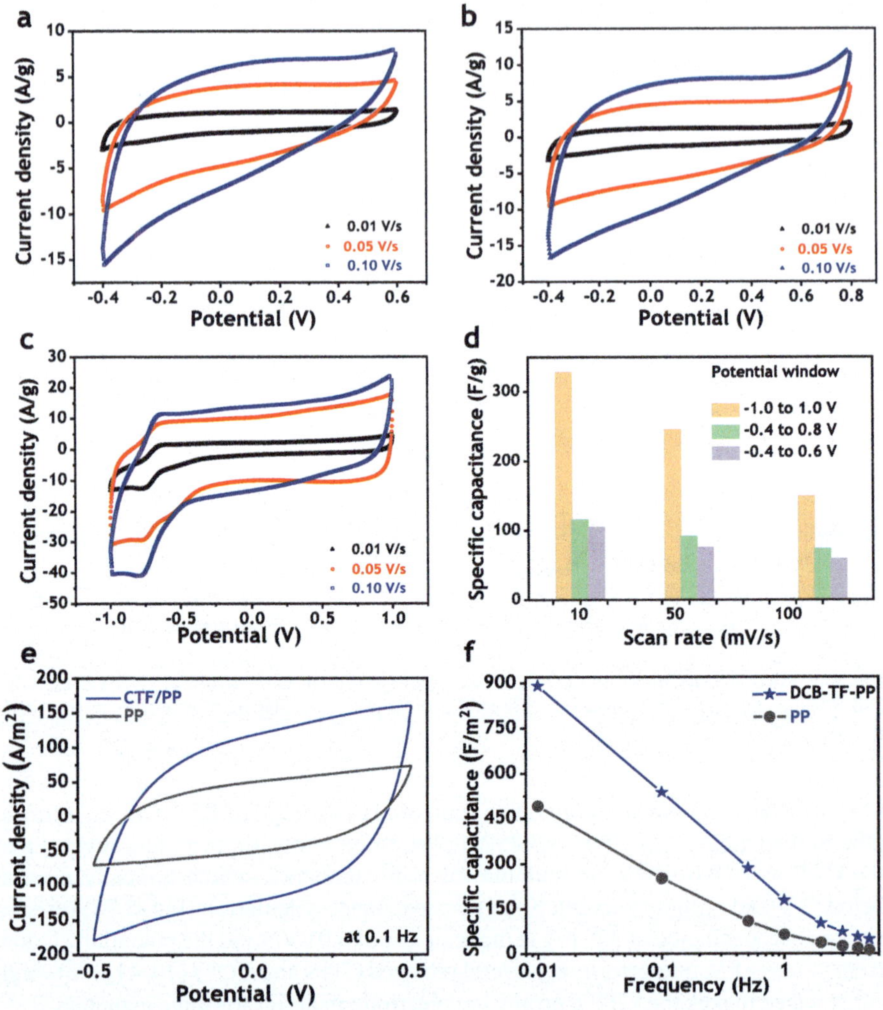

Fig. 5.7 Specific capacitance [29] from Ref. [29]. Reprinted with permission from John Wiley and Sons: **a–c** CVs at different potential windows and **d** specific capacitances of CTF; **e** CVs and **f** specific capacitances of CTF/PP electrode in actuator

5.3.4 Actuation Performance

To further evaluate its performance, a CTF/PP (4 wt.%) electrode mixture was cast on a Nafion membrane, creating a CTF/PP actuator with a thickness of about 135 μm. For comparison, a PP actuator without CTF was prepared following the same procedure. These actuators, with a free length of 2 cm, were tested under DC and AC voltages, and the actuation performance is displayed in Fig. 5.8.

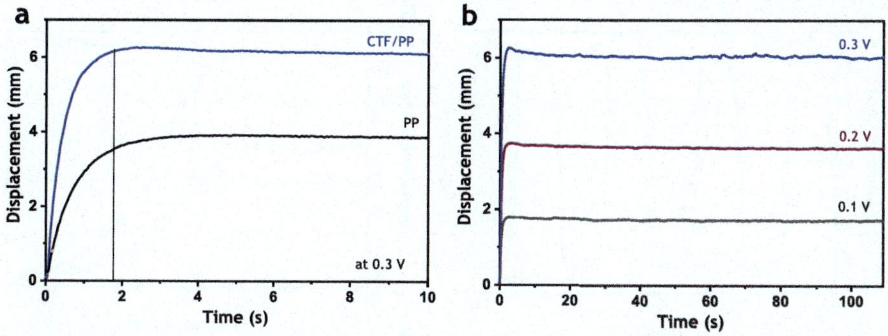

Fig. 5.8 DC response [29] from Ref. [29]. Reprinted with permission from John Wiley and Sons: **a** rise time of CTF/PP and PP actuators; **b** responses of CTF/PP actuator

The DC responses at 0.3 V (Fig. 5.8a) shows that while the addition of CTF did not improve the response time (approximately 2 s), it enhanced bending deformation by over 1.5 times, from under 4 mm to more than 6 mm. Figure 5.8b demonstrates the absence of back relaxation for over 100 s at different voltages, with the deformation increasing linearly from under 2 mm to over 6 mm as voltage amplitudes ranged from 0.1 to 0.3 V. This enhanced performance suggests that the CTF/PP actuator's increased capacitance, owing to the porous CTF, enables greater bending at the same voltage.

AC excitation testing revealed further peformance improvement (Fig. 5.9). At 0.3 V and 0.1 Hz, the CTF/PP actuator exhibited a bending displacement of over 13 mm, compared to under 8 mm for the PP actuator (Fig. 5.9a). This improvement doubled at 0.5 V and 0.1 Hz (Fig. 5.9c) and tripled at 1.0 V and 2.0 Hz (Fig. 5.9e). These enhancements were consistent across a broad frequency range (0.01 to 20.00 Hz at 0.3 V in Fig. 5.9b and 0.5 V in Fig. 5.9d) and varying voltages (0.01–1.00 V at 2 Hz in Fig. 5.9e). These results suggest that while the response pattern, determined by ionic resistance in electrolytes, remained unchanged, the presence of CTF inhanced performance by generating stronger electrical fields.

The actuator's stability and blocking forces were tested (Fig. 5.9f–g). The stabililty was evaluated under continuous 0.5 V at 1 Hz, with normalized displacements calculated relative to the average over 39 days. The actuator maintained remarkable stability over 39 days, showing minimal degradation after more than 3.4 million cycles (Fig. 5.9f). This durability is attributed to the high-quality of the developed CTF and Nafion membrane, which enebles efficient charge and ion transfer, maintaining the interface bonding between the electrode and the electrolyte. The blocking forces in Fig. 5.9g revealed three distinct regions in the voltage-force relationship. From 0.1 to 0.6 V, the force increased rapidly and linearly, slowing between 0.7 and 1.5 V, and stabilizing beyond 1.5 V. Compared to its PP counterpart, the CTF/PP actuator showed twofold increase in forces, normalized values of 11.7 and 22.9 at 1.5 V, respectively. This enhanced performance is attributed to the superior properties of the CTF material.

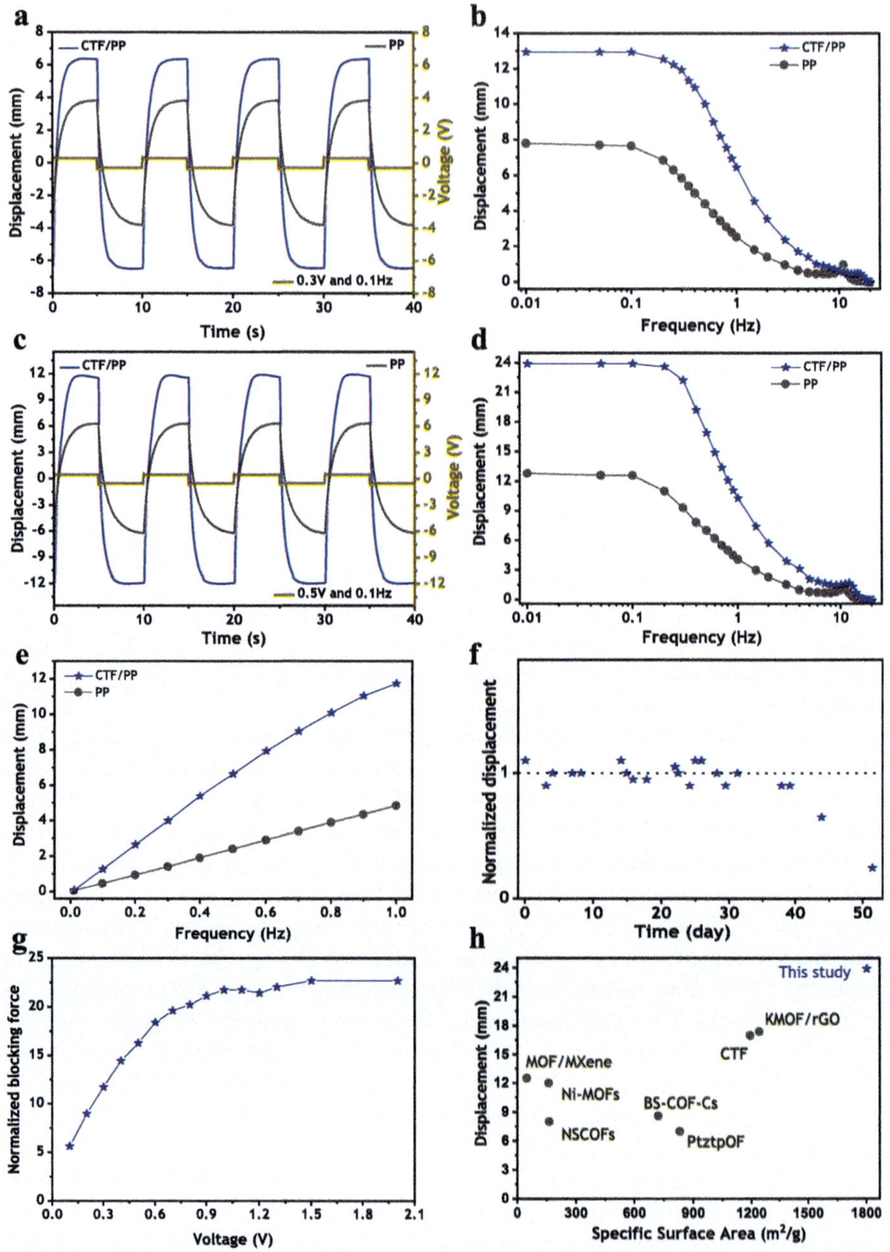

Fig. 5.9 Performance [29] from Ref. [29]. Reprinted with permission from John Wiley and Sons: AC responses at **a** 0.3 V and 0.1 Hz, **b** 0.3 V with different frequency, **c** 0.5 V and 0.1 Hz, **d** 0.5 V with different frequency, **e** 2 Hz with different voltage; **f** durability; **g** blocking forece; **h** comparison, MOF/MXene [35], Ni-MOFs [10], NSCOFs [4], BS-COF-Cs [23], PztzpOF [22], CTF [6], KMOF/rGO [34]

The superior performance of the CTF, compared to other MOFs and COFs used in ionic actuators, is clearly demonstrated by comparing specific surface areas and actuator displacements (Fig. 5.8h). Previously reported materials had surface areas between 50 and 1190 m^2/g and displacement values between 7 and 17 mm at 0.5 V and 0.1 Hz. In contrast, the CTF exhibited a surface area of 1800 m^2/g and a displacement of 24 mm, outperforming the highest reported values for both surface area and bending deformation, including the top-performing CTF [6] and KMOF/rGO [34]. This comparison validates the bone-inspired approach for designing CTF electrodes and highlights the critical role of surface area and porosity in optimizing actuation performance.

5.3.5 Demonstration

This bone-inspired CTF electrode significantly advanced the performance of ionic actuators, making them well-suited for applications in some soft robots. For example, this actuator was used to flap dragonfly wings in a kinetic art installation (Fig. 5.10a). The setup involved five dragonflies perched on flower petals, with actuators attached to their wings. The electrical connection, established using cupper tape and steel screws, successfully powered forty electrodes of twenty actuators, which in turn moved twenty plastic wings. The entire structure was mounted on a rotating motor. Static tests (Fig. 5.10b) showed that these small actuators could produce large wing-flapping at 1 V. Even when the flower was rotated (Fig. 5.10c), the actuators maintained continuous wing flapping, demonstrating their robustness. This complex demonstration underscores the high quality of the bone-inspired CTF electrode material and the potential of its actuators for further applications (Fig. 5.10).

5.4 Conclusion

This chapter presents the successful development of a bone-inspired CTF derived from 2,2′-bipyridine-6,6′-dicarbonitrile as a promising electrode material for ionic actuators. Synthesizing using an ionothermal technique at 600 °C, the CTF exhibited the five key properties essential for ionic actuators: a large specific surface area of 1,800 m^2/g, hierarchical porosity, high stability, a high nitrogen content, and good conductivity due to partial carbonization and conjugated pi-bonds of large aromatic ring system. These properties were confirmed through various characterization techniques, including TEM, SEM, elemental mapping, FT-IR, solid-state ^{13}C NMR, XPS, argon adsorption–desorption isotherm, Raman spectroscopy, TGA, and CV. With a high specific capacitance of 330 F/g, this CTF significantly improved actuation performance across a wide range of voltages (0.01–1.00 V) and frequencies (0.01–20.00 Hz). Particularly doubling the bending displacement to 24 mm at 0.5 V and 0.2 Hz. This exceptional performance, coupled with the successful kinetic

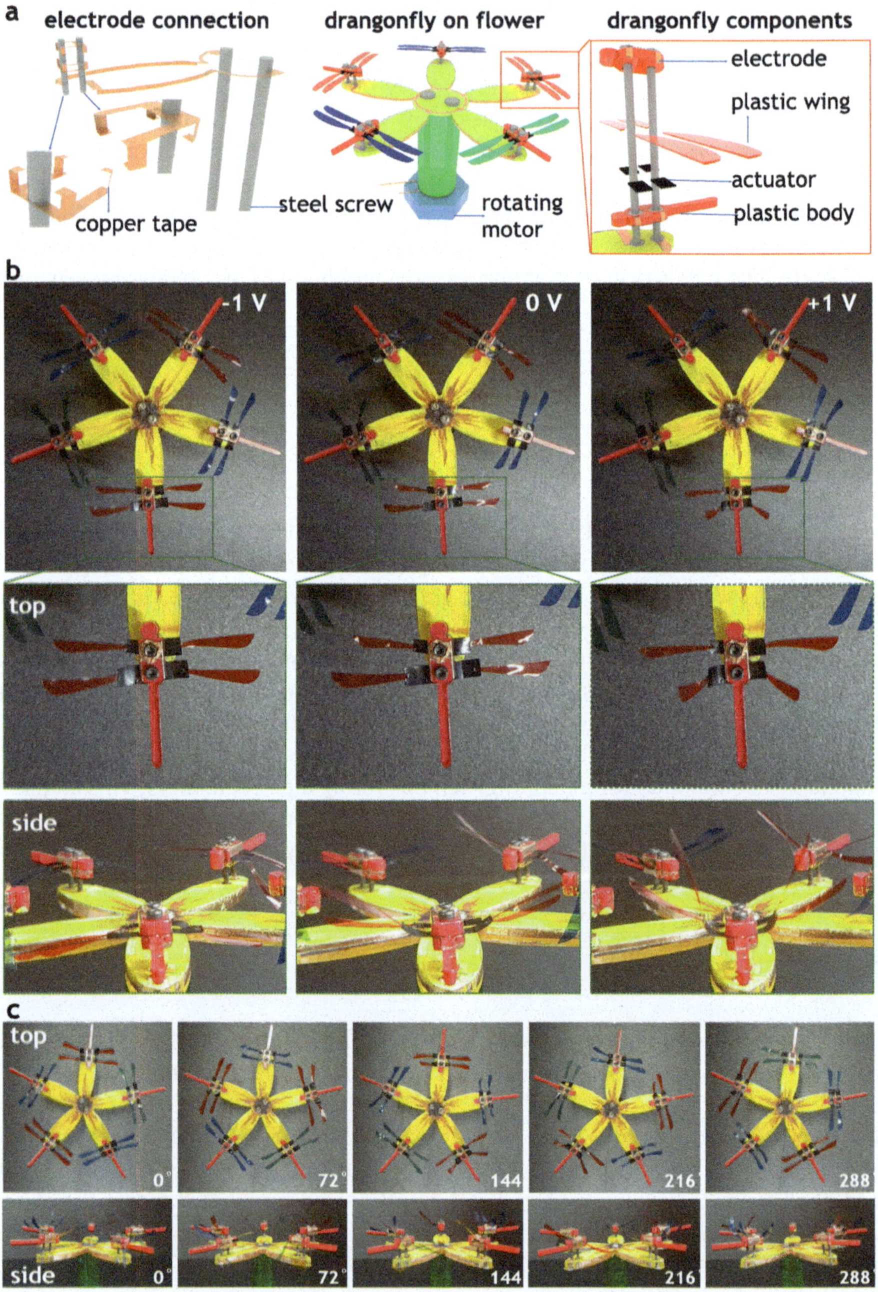

Fig. 5.10 Dragonfly kinetic art [29] from Ref. [29]. Reprinted with permission from John Wiley and Sons: **a** scheme of actuator used in dragonflies; **b** flapping wings; **c** rotating flower

art demonstration, validates the bone-inspired design approach and highlights the outstanding potential of this CTF for future applications.

References

1. Yeom SW, Oh IK (2009) A biomimetic jellyfish robot based on ionic polymer metal composite actuators. Smart Mater Struct 18. https://doi.org/10.1088/0964-1726/18/8/085002
2. Must I et al (2015) Ionic and capacitive artificial muscle for biomimetic soft robotics. Adv Eng Mater 17:84–94. https://doi.org/10.1002/adem.201400246
3. Manzoor MT et al (2019) Mutually exclusive p-type and n-type hybrid electrode of MoS_2 and graphene for artificial soft touch fingers. Adv Funct Mater. https://doi.org/10.1002/adfm.201905454
4. Mahato M et al (2020) Sulfur- and nitrogen-rich porous π-conjugated COFs as stable electrode materials for electro-ionic soft actuators. Adv Funct Mater 30. https://doi.org/10.1002/adfm.202003863
5. Mahato M et al (2020) Covalent organic frameworks: sulfur- and nitrogen-rich porous π-conjugated COFs as stable electrode materials for electro-ionic soft actuators. Adv Funct Mater 30(46). https://doi.org/10.1002/adfm.202070300
6. Mahato M et al (2020) CTF-based soft touch actuator for playing electronic piano. Nat Commun 11:5358. https://doi.org/10.1038/s41467-020-19180-3
7. Tabassian R et al (2019) Graphene mesh for self-sensing ionic soft actuator inspired from mechanoreceptors in human body. Adv Sci. https://doi.org/10.1002/advs.201901711
8. Tabassian R et al (2019) Self-sensing actuators: graphene mesh for self-sensing ionic soft actuator inspired from mechanoreceptors in human body. Adv Sci 6(23). https://doi.org/10.1002/advs.201970137
9. Umrao S et al (2019) MXene artificial muscles based on ionically cross-linked $Ti_3C_2T_x$ electrode for kinetic soft robotics. Sci Robot 4. https://doi.org/10.1126/scirobotics.aaw7797
10. Mahato M et al (2022) A dual-responsive magnetoactive and electro-ionic soft actuator derived from a nickel-based metal-organic framework. Adv Mater 34:e2203613. https://doi.org/10.1002/adma.202203613
11. Asaka K, Oguro K, Nishimura Y, Mizuhata M, Takenaka H (1995) Bending of polyelectrolyte membrane–platinum composites by electric stimuli I. Response characteristics to various waveforms. Polym J 27:436–440. https://doi.org/10.1295/polymj.27.436
12. Oguro K, Fujiwara N, Asaka K, Onishi K, Sewa S (1999) Polymer electrolyte actuator with gold electrodes. Proc Spie 3669:64–71. https://doi.org/10.1117/12.349698
13. Vargantwar PH, Shankar R, Krishnan AS, Ghosh TK, Spontak RJ (2011) Exceptional versatility of solvated block copolymer/ionomer networks as electroactive polymers. Soft Matter 7:1651. https://doi.org/10.1039/c0sm01210f
14. Zhou DZ et al (2003) Solid state actuators based on polypyrrole and polymer-in-ionic liquid electrolytes. Electrochim Acta 48:2355–2359. https://doi.org/10.1016/S0013-4686(03)00225-1
15. Kim SS, Jeon JH, Kee CD, Oh IK (2013) Electro-active hybrid actuators based on freeze-dried bacterial cellulose and PEDOT:PSS. Smart Mater Struct 22. https://doi.org/10.1088/0964-1726/22/8/085026
16. Fukushima T, Asaka K, Kosaka A, Aida T (2005) Fully plastic actuator through layer-by-layer casting with ionic-liquid-based bucky gel. Angew Chem Int Ed Engl 44:2410–2413. https://doi.org/10.1002/anie.200462318
17. Mukai K et al (2009) Highly conductive sheets from millimeter-long single-walled carbon nanotubes and ionic liquids: application to fast-moving, low-voltage electromechanical actuators operable in air. Adv Mater 21:1582–1585. https://doi.org/10.1002/adma.200802817

18. Kotal M, Kim J, Kim KJ, Oh IK (2016) Sulfur and nitrogen co-doped graphene electrodes for high-performance ionic artificial muscles. Adv Mater 28:1610–1615. https://doi.org/10.1002/adma.201505243
19. Kim J, Jeon JH, Kim HJ, Lim H, Oh IK (2014) Durable and water-floatable ionic polymer actuator with hydrophobic and asymmetrically laser-scribed reduced graphene oxide paper electrodes. ACS Nano 8:2986–2997. https://doi.org/10.1021/nn500283q
20. Wu G et al (2015) Graphitic carbon nitride nanosheet electrode-based high-performance ionic actuator. Nat Commun 6:7258. https://doi.org/10.1038/ncomms8258
21. Kotal M et al (2018) Highly bendable ionic soft actuator based on nitrogen-enriched 3D hetero-nanostructure electrode. Adv Funct Mater 28. https://doi.org/10.1002/adfm.201802464
22. Roy S, Kim J, Kotal M, Kim KJ, Oh IK (2017) Electroionic antagonistic muscles based on nitrogen-doped carbons derived from poly(Triazine-Triptycene). Adv Sci (Weinh) 4:1700410. https://doi.org/10.1002/advs.201700410
23. Roy S et al (2019) Collectively exhaustive electrodes based on covalent organic framework and antagonistic co-doping for electroactive ionic artificial muscles. Adv Funct Mater. https://doi.org/10.1002/adfm.201900161
24. Ahn T et al (2021) Hierarchical nature of nanoscale porosity in bone revealed by positron annihilation lifetime spectroscopy. ACS Nano 15:4321–4334. https://doi.org/10.1021/acsnano.0c07478
25. Buckwalter JA, Glimcher MJ, Cooper RR, Recker R (1995) Bone biology. J Bone Jt Surg 77:1256–1275. https://doi.org/10.2106/00004623-199508000-00019
26. Wegst UG, Bai H, Saiz E, Tomsia AP, Ritchie RO (2015) Bioinspired structural materials. Nat Mater 14:23–36. https://doi.org/10.1038/nmat4089
27. Weiner S, Wagner HD (1998) The material bone: structure mechanical function relations. Annu Rev Mater Sci 28:271–298. https://doi.org/10.1146/annurev.matsci.28.1.271
28. Grandfield K, Vuong V, Schwarcz HP (2018) Ultrastructure of bone: hierarchical features from nanometer to micrometer scale revealed in focused ion beam sections in the TEM. Calcif Tissue Int 103:606–616. https://doi.org/10.1007/s00223-018-0454-9
29. Garai M et al (2024) Multifunctional and electronically conjugated triazine framework for superior electro-ionic artificial muscles. Adv Funct Mater. https://doi.org/10.1002/adfm.202406603
30. Feng X, Ding X, Jiang D (2012) Covalent organic frameworks. Chem Soc Rev 41:6010–6022. https://doi.org/10.1039/c2cs35157a
31. James SL (2003) Metal-organic frameworks. Chem Soc Rev 32:276–288. https://doi.org/10.1039/b200393g
32. Zhou HC, Long JR, Yaghi OM (2012) Introduction to metal-organic frameworks. Chem Rev 112:673–674. https://doi.org/10.1021/cr300014x
33. Zhou HC, Kitagawa S (2014) Metal-organic frameworks (MOFs). Chem Soc Rev 43:5415–5418. https://doi.org/10.1039/c4cs90059f
34. Kotal M, Tabassian R, Roy S, Oh S, Oh IK (2020) Metal-organic framework-derived graphitic nanoribbons anchored on graphene for electroionic artificial muscles. Adv Funct Mater. https://doi.org/10.1002/adfm.201910326
35. Garai M et al (2022) Metal organic framework-MXene nanoarchitecture for fast responsive and ultra-stable electro-ionic artificial muscles. Adv Funct Mater. https://doi.org/10.1002/adfm.202212252
36. Nguyen VH et al (2024) Functionally antagonistic polyelectrolyte for electro-ionic soft actuator. Nat Commun 15:435. https://doi.org/10.1038/s41467-024-44719-z
37. Katekomol P, Roeser J, Bojdys M, Weber J, Thomas A (2013) Covalent triazine frameworks prepared from 1,3,5-tricyanobenzene. Chem Mater 25:1542–1548. https://doi.org/10.1021/cm303751n
38. Ferrari AC, Robertson J (2000) Interpretation of Raman spectra of disordered and amorphous carbon. Phys Rev B 61:14095–14107. https://doi.org/10.1103/PhysRevB.61.14095
39. Guo J et al (2020) High voltage supercapacitor based on nonflammable high-concentration-ionic liquid electrolyte. Colloids Surf A: PhysChemical Eng Asp 598. https://doi.org/10.1016/j.colsurfa.2020.124858

40. Kandambeth S, Kale VS, Shekhah O, Alshareef HN, Eddaoudi M (2022) 2D Covalent-organic framework electrodes for supercapacitors and rechargeable metal-ion batteries. Adv Energy Mater 12. https://doi.org/10.1002/aenm.202100177
41. Liu X, Liu CF, Lai WY, Huang W (2020) Porous organic polymers as promising electrode materials for energy storage devices. Adv Mater Technol-Us 5. https://doi.org/10.1002/admt.202000154

Chapter 6
Concluding Remark

This dissertation, including an introductory chapter, four main chapters detailing each study, and a concluding chapter, presented four interdisciplinary studies focused on developing bioinspired nanomaterials with novel combinations of properties for use in ionic artificial muscles. Beyond nanomaterials and ionic artificial muscles, the research spanned polymer synthesis, micelle science, nanocomposite, CTF synthesis, electrochemistry, bioinspiration and biomimetics, design mechanics, and soft robotics.

Chapter 1 provided a comprehensive literature review, summarizing and analyzing major materials used in ionic artificial muscles. It highlighted that while electrolytes need high ionic conductivity and good mechanical strength, electrodes require high electrical conductivity and large capacitance. These properties often have an inverse relationship, posing significant challenges in material and technology development. Despite numerous advancements, this issue remains unsresolved, warranting further research.

The review also indicated promising materials for ionic artificial muscles, such as block isomers and Nafion for electrolytes, nanocomposites and CTFs for electrodes. Inspired by nature, this dissertation developed three materials (a block isomer, a nanocomposite, and a CTF) and one technology, addressing the inverse relationship. The block isomer and Nafion featured unique micellar structures, providing electrolyte membranes with high ionic conductivity and mechanical strength. The nanocomposite combined graphene for conductivity and MoS_2 for charge storage, creating electrodes with both desirable properties. The CTF, with its unique structure and composition, generated electrodes with high capacitance and good electrical conductivity.

Chapter 2 discussed the block isomer, PS-b-PSS-EMIm, inspired by cell membrane. Its successful synthesis in core–shell structure was confirmed by GPC, ^{1}H NMR, SEM, TEM. These unique micelles formed micro-thick membranes with high ionic conductivity and good mechanical strength, as evidenced by tensile tests,

V. H. Nguyen, *Bioinspired Multifunctional Nanomaterials for Ionic Artificial Muscles*, Springer Theses, https://doi.org/10.1007/978-3-031-78813-0_6

DSC, and EIS data. The resulting actuator achieved over 8 mm displacement at 0.5 V and 0.1 Hz, with a notably faster response time compared to other block isomers.

Chapter 3 detailed the preparation of Nafion membranes inspired by hemostasis. Nafion micelles, with non-conducting core and sulfonate group shell, were prepared in a water-DMAc mixture and characterized by cryo-TEM, DLS, and zeta potential. These micelles produced membranes with continuous conducting phase and well-connected non-conducting domains, shown by tensile test, CV, and EIS. These membranes exhibited twice the ionic conductivity while maintaining mechanical strength, resulting in over 35 times improved bending deformation at 1 V and 3 Hz, suitable for inchworm-mimetic robots and kinetic tensegrity systems.

Chapter 4 introduced a nanocomposite, inspired by calcium metabolism. MoS_2, with high specific capacitance, and graphene, with excellent conductivity, were selected to develop the nanocomposite. Additionally, these two-dimensional materials have good stability and high surface areas. Direct growth of MoS_2 on graphene enhanced bonding and charge transfer. This bioinspired material increased actuation bending by 6.5 times compared to rGO at 0.5 V and 1 Hz, enabling the actuators to function like soft fingers interacting with delicate display surfaces.

Chapter 5 explored the BPDC-derived CTF, inspired by bone porous structure. Synthesized by an ionothermal method and characterized by SEM, TEM, FT-IR, solid-state ^{13}C NMR, XPS, Raman spectroscopy, TGA, argon adsorption–desorption isotherms, and CV, this CTF exhibited a high specific capacitance of 330 F/g due to its large specific area (1,800 m^2/g), high nitrogen-heteroatom percentage, and good conductivity. This resulted in over 2.5 times greater bending deformation at 0.5 V and 1.0 Hz, surpassing other reported COFs and MOFs. Therefore, the derived actuators were demonstrated to operate dragonfly wings.

Overall, this interdisciplinary research advances material science and ionic actuator engineering while impacting related fields such as chemistry, electrochemistry, bioinspiration, biomimetics, design mechanics, and soft robotics. The extensive literature analysis identified key requirements and promising materials for ionic actuators. The discoveries of bioinspired materials, including one block isomer, one nanocomposite electrode, and one CTF, and one technology for micro-thick electrolyte membranes, contribute significantly to nanomaterials and encourage nature-inspired innovations. The enhanced actuation performance accelerates the growth of ionic actuators. The successful demonstrations promote the development of soft robotics. Furthermore, the methodologies and materials developed here may extend to fuel cells, supercapacitors, batteries, making a significant step forward in these fields.

Curriculum Vitae

Van Hiep Nguyen

vhnguyen1003@kaist.ac.kr

https://orcid.org/0000-0003-4180-2352

https://scholar.google.com/citations?user=Rxd3deIAAAAJ&hl=en.

Education

2018–2023 Ph.D. Mechanical Engineering, Korea Advanced Institute of Science and Technology (KAIST), Korea.

2015–2018 M.S. Mechanical Engineering, KAIST, Korea.

2003–2008 B.S. Polymer and composite materials, Hanoi University of Science and Technology (HUST), Vietnam.

Research Interests

Material Synthesis: Polymer electrolytes, COFs, MOFs, Graphene, MXenes, Bioinspired nanomaterials.

Device Applications: Actuators, Soft robotics, Energy conversion and storage.

Research Experience

- Developing magnetic materials derived from iron MOFs and solid electrolytes for actuators and bioapplications
- Developed a procedure for controlling ionic conducting nanochannels of Nafion membrane
- Developed a covalent triazine framework (CTF) with high surface area and rich nitrogen content
- Prepared graphene oxide (GO), reduced GO (rGO), doped rGO, and MoS_2/rGO

V. H. Nguyen, *Bioinspired Multifunctional Nanomaterials for Ionic Artificial Muscles*, Springer Theses, https://doi.org/10.1007/978-3-031-78813-0

- Prepared ionic block copolymer electrolytes by RAFT polymerization
- Fabricated and characterized ionic actuators and soft robotic demonstrations
- Contributed to projects about graphene, COFs, MOFs, polymers, Mxenes, etc. for actuators, batteries, capacitors
- Attended Elsevier Researcher Academy and MXene course by professor Yury Gogotsi's group
- Manufactured and characterized composite materials via vacuum assisted resin transfer molding
- Synthesized polyimides, polyesters, cresol formaldehyde, epoxy esters, polymer electrolytes
- Manufactured and characterized some paints especially electrophoretic deposition
- Prepared and characterized natural fibers and their green composites.

Peer-Reviewed Papers During Graduate Study

1. M. Kotal, J. Kim, R. Tabassian, S. Roy, V. H. Nguyen, N. Koratkar, I.-K. Oh*, "Highly Bendable Ionic Soft Actuator Based on Nitrogen-Enriched 3D Hetero-Nanostructure Electrode", Advanced Functional Materials 1,802,464, 2018.
2. R. Tabassian, J. Kim, V. H. Nguyen, M. Kotal, I.-K. Oh, "Functionally Antagonistic Hybrid Electrode with Hollow Tubular Graphene Mesh and Nitrogen-Doped Crumpled Graphene for High-Performance Ionic Soft Actuators", Advanced Functional Materials 1,705,714, 2018.
3. J. Kim, R. Tabassian, V. H. Nguyen, S. Umrao, I.-K. Oh, "Crumpled quaternary nano-architecture of sulfur-doped nickel cobalt selenide directly grown on carbon cloth for making stronger ionic soft actuators", ACS Applied Materials & Interfaces 40,451–40,460, 2019.
4. S. Umrao, R. Tabassian, J. Kim, V. H. Nguyen, Q. Zhou, S. Nam, I.-K. Oh, "MXene artificial muscles based on ionically cross-linked $Ti_3C_2T_x$ electrode for kinetic soft robotics", Science Robotics eaaw7797, 2019.
5. R. Tabassian, V. H. Nguyen, S. Umrao, M. Mahato, J. Kim, M. Porfiri, I.-K. Oh*, "Graphene mesh for self-sensing ionic soft actuator inspired from mechanoreceptors in human body", Advanced Science 1,901,711, 2019.
6. M. T. Manzoor, V. H. Nguyen, S. Umrao, J. Kim, R. Tabassian, J. E. Kim, I.-K. Oh, "Mutually exclusive p-type and n-type hybrid electrode of MoS_2 and graphene for artificial soft touch fingers", Advanced Functional Materials 1,905,454, 2019, M. T. M. and V. H. N. equal contribution.
7. V. H. Nguyen, J. Kim, R. Tabassian, M. Kotal, K. Jun, J.-H. Oh, J.-M. Son, M. T. Manzoor, K. J. Kim, I.-K. Oh*, "Electroactive Artificial Muscles Based on Functionally Antagonistic Core–shell Polymer Electrolyte Derived from PS-B-PSS Block Copolymer", Advanced Science 1,801,196, 2019.
8. J.-H. Oh, J.-S. Kim, V. H. Nguyen, I.-K. Oh*, "Auxetic Graphene Oxide-Porous Foam for Acoustic Wave and Shock Energy Dissipation", Composites Part B 107,817, 2020.

9. M. Mahato, R. Tabassian, V. H. Nguyen, S. Oh, S. Nam, K. J. Kim, I.-K. Oh*, "Sulfur- and nitrogen-rich porous π-conjugated COFs as stable electrode materials for electro-ionic soft actuators", Advanced Functional Materials 2,003,836, 2020.
10. S. Oh, V. H. Nguyen, V. T. Bui, S. Nam, M. Mahato, I.-K. Oh*, "Intertwined nanosponge solid-state polymer electrolyte for rollable and foldable lithium-ion batteries", ACS Applied Materials & Interfaces 11657–11668, 2020.
11. M. Mahato, R. Tabassian, V. H. Nguyen, S. Oh, S. Nam, W.-J. Hwang, I.-K. Oh*, "CTF-based soft touch actuator for playing electronic piano", Nature Communications 5358, 2020.
12. V. H. Nguyen, R. Tabassian, S. Oh, S. Nam, M. Mahato, P. Thangasamy, a. Rajabi-Abhari, W.-J. Hwang, a. K. Taseer, I.-K. Oh*, "Stimuli-Responsive MXene-Based Actuators", Advanced Functional Materials 1,909,504, 2020.
13. S. Nam, J. Kim, V. H. Nguyen, M. Mahato, S. Oh, P. Thangasamy, C. W. Ahn, I.-K. Oh*, "Collectively exhaustive MXene and graphene oxide multilayer for suppressing shuttling effect in flexible lithium sulfur battery", Advanced Materials Technologies 2,101,025, 2021.
14. R. Tabassian, M. Mahato, S. Nam, V. H. Nguyen, a. Rajabi-Abhari, I.-K. Oh*, "Electro-Active and Photo-Active Vanadium Oxide Nanowire Thermo-Hygroscopic Actuators for Kirigami Pop-Up", Advanced Science 2,102,064, 2021.
15. M. Mahato, S. Nam, R. Tabassian, S. Oh, V. H. Nguyen, I.-K. Oh*, "Electronically conjugated multifunctional covalent triazine framework for unprecedented CO_2 selectivity and high-power flexible supercapacitor", Advanced Functional Materials 2,107,442, 2021.
16. S. Oh, R. Tabassian, P. Thangasamy, M. Mahato, V. H. Nguyen, S. Nam, Z. Huapeng, I.-K. Oh*, "Cooling-accelerated nanowire-nitinol hybrid muscle for versatile prosthetic hand and biomimetic retractable claw", Advanced Functional Materials 2,111,145, 2021.
17. U. Raza, S. Oh, R. Tabassian, M. Mahato, V. H. Nguyen, I.-K. Oh*, "Micro-Structured Porous Electrolytes for Highly Responsive Ionic Soft Actuators", Sensors & Actuators: B. Chemical 131,006, 2021.
18. M. Mahato, W.-J. Hwang, R. Tabassian, S. Oh, V. H. Nguyen, S. Nam, J.-S. Kim, H. Yoo, A. K. Taseer, M.-J. Lee, H. Zhang, T.-E. Song, I.-K. Oh*, "A dual-responsive magnetoactive and electro-ionic soft actuator derived from a nickel-based metal-organic framework", Advanced Materials 2203613, 2022.
19. M. Garai, M. Mahato, S. Nam, E. Kim, D. Seo, Y. Lee, V. H. Nguyen, S. Oh, P. Sambyal, H. Yoo, a. K. Taseer, S. S. Ali, H. Han, C. W. Ahn, J. Kim, I.-K. Oh*, "Metal Organic Framework-MXene Nanoarchitecture for Fast Responsive and Ultra-Stable Electro-Ionic Artificial Muscles", Advanced Functional Materials 2,212,252, 2023.
20. P. Sambyal, M. Mahato, a. K. Taseer, H. Yoo, M. Garai, V. H. Nguyen, S. S. Ali, I.-K. Oh*, "Magnetically and Electrically Responsive Soft Actuator Derived from Ferromagnetic Bimetallic Organic Framework", Small 2,207,140, 2023.

21. M. Mahato, M. Garai, V. H. Nguyen, S. Oh, S. Nam, X. Zeng, H. Yoo, R. Tabassian*, I.-K. Oh*, "Polysulfonated Covalent Organic Framework as Active Electrode Host for Mobile Cation Guests in Electrochemical Soft Actuator", Science Advances, Eadk9752, 2023.
22. S. S. Ali, M. Mahato, D. V. Lam, P. Sambyal, G. Valurouthu, M. Garai, a. Dewan, V. H. Nguyen, M. Khan, a. K. Taseer, I.-K. Oh*, "Mxene-Cobalt Hbrid Electrodes for Electroactive Artificial Muscle", Advanced Engineering Materials, 2,400,515, 2024.
23. X. Zeng, M. Mahato, W. Oh, H. Yoo, V. H. Nguyen, S. Oh, G. Valurouthu, S.-K. Jeong, C. W. Ahn, Y. Gogotsi, I.-K. Oh*, "Stoichiometric $Ti_3C_2T_x$ Coating for Inhibiting Dendrite Growth in Anode-Free Lithium Metal Batteries", Energy and Environmental Materials, E12686, 2024.
24. H. Yoo, M. Mahato, J.-S. Kim, S. Oh, M. Garai, V. H. Nguyen, a. K. Taseer, M.-J. Lee, I.-K. Oh*, "Janus CoMOF-SEBS Membrane for Bifunctional Dielectric Layer in Triboelectric Nanogenerators", Advanced Science, 2,307,656, 2024.
25. A. K. Taseer, S. Oh, J.-S. Kim, M. Garai, H. Oh, V. H. Nguyen, M. Mahato*, I.-K. Oh*, "Cobalt MOF-Based Porous Carbonaceous Spheres for Multimodal Soft Actuator Exhibiting Intricate Biomimetic Motions", Advanced Materials, 2,312,340, 2024.
26. M. Garai, V. H. Nguyen, M. Mahato, S. Oh, D. Saatchi, H. Yoo, S. S. Ali, D. V. Lam, A. K. Taseer, I.-K. Oh*, "Multifunctional and electronically conjugated triazine framework for superior electro-ionic artificial muscles", Advanced Functional Materials, 2,406,603, 2024, M.G. and V. H. N. equal contribution.
27. V. H. Nguyen, S. Oh, M. Mahato, R. Tabassian, H. Yoo, S.-G. Lee, M. Garai, K. J. Kim, I.-K. Oh*, "Functionally Antagonistic Polyelectrolyte for Electro-Ionic Soft Actuator", Nature Communications 435, 2024.

Oral Conference Presentations During Graduate Study

1. V. H. Nguyen, J. Kim, R. Tabassian, M. Kotal, I.-K. Oh*, "Polystyrene-B-Poly(1-Hexyl-3-Methylimidazolium 4-styrenesulfonate) Block Copolymer for Ionic Artificial Muscle", the 5th Annual Conference of Vietnamese Young Scientists, Sejong University, Seoul, Korea, May 20, 2018.
2. V. H. Nguyen, J. Kim, I.-K. Oh*, W. J. Kim, "Design and synthesis of PS-b-PSS block copolymer electrolytes for electroactive artificial muscles", The 3rd international workshop on active materials and soft mechatronics, KAIST, Daejeon, Korea, October 23–26, 2018.
3. V. H. Nguyen, I.-K. Oh*, "Block copolymer electrolyte for ionic electroactive artificial muscles", SPIE Smart Structure + Nondestructive Evaluation, 115870 V, March 22, 2021, virtual.
4. V. H. Nguyen, I.-K. Oh*, "Nanocomposite electrodes for high-performance ionic artificial muscles", The Korean society for composite materials (KSCM2021), Korea, November 11–12, 2021, virtual.
5. V. H. Nguyen, M. T. Manzoor, I.-K. Oh*, "Molybdenum disulfide/graphene nanocomposite as electrodes for ionic artificial muscles", SPIE Smart Structure + Nondestructive Evaluation on Demand, March 6 - 9, 2022, virtual.

6. V. H. Nguyen, I.-K. Oh*, "Block copolymer by solution RAFT polymerization for ionic electroactive artificial muscles", The Korean society for composite materials (KSCM2022), Korea, May 18–20, 2022, virtual.
7. V. H. Nguyen, I.-K. Oh*, "Continuous conducting phases of electrolyte membranes booming performance of ionic artificial muscles", The 8th annual conference of Vietnamese young scientists, Seoul National University of Science and Technology, Seoul, Korea, December 11, 2022.
8. V. H. Nguyen, I.-K. Oh*, "Guidelines for developing electrode and polymer electrolyte nanomaterials with desirable properties for ionic actuators", The 8th annual conference of Vietnamese young scientists, Seoul National University of Science and Technology, Seoul, Korea, December 11, 2022.
9. V. H. Nguyen, I.-K. Oh*, "Developing a novel block isomer for ionic actuators", The Polymer society of Korea (PSOK2023), Daejeon Convention Center, Daejeon, Korea, April 05–07, 2023.
10. V. H. Nguyen, M. Garai, I.-K. Oh*, "A Novel Covalent Triazine Framework Electrode for Ionic Actuator", The 9th annual conference of Vietnamese young scientists, Sejong University, Seoul, Korea, July 16, 2023.
11. V. H. Nguyen, I.-K. Oh*, "Micellar assembly intensifying performance of electroactive ionic actuators", The Korean society for composite materials (KSCM2024), Jeju Shinhwa World, Korea, April 24–26, 2024.

Service Activities

- Supported Vice Head of Department of Mechanical Engineering, KAIST to connect and visit VinUniversity, HUST, Vietnam National University – University of Engineering and Technology, 2024
- Supported mutual visits bewteen KAIST and Vietnamese faculties, 2017, 2023–2024
- Visited and presented to HUST students about KAIST graduate admission, 2023
- Introducing KAIST to Vietnamese students, guiding admission process and their beginning at KAIST, 2018–2024
- A leader of International Student Organization, Mechanical Engineering Department, KAIST, 2023–2024
- KAIST Ambassador, led an online meeting presenting KAIST to Vietnamese students, 2023
- Representative for international married students, KAIST dormitory, 2020–2022.

Scholarships and Awards

2018	Best Presentation Award, the 5th annual conference of Vietnamese young scientists.
2015–2022	KAIST Scholarship, waiving tuition fees for 6 years during master and doctoral study.
2015–2022	Brain Korea 21 Scholarship, 6 years during master and doctoral study.
2003–2008	HUST Scholarship for excellent academic performance, 2nd, 4th, 6–9th semesters.

References

1. Prof. Dr. Oh, Il-Kwon (master, Ph.D., and postdoc advisor), Department of Mechanical Engineering, KAIST, Korea, E-Mail: ikoh@kaist.ac.kr, Phone: + 82 42 350 1520.
2. Assoc. Prof. Dr. Nguyen, Huy Tung (B.S. thesis advisor), School of Materials Science and Engineering, HUST, Vietnam, E-mail: tung.nguyenhuy@hust.edu.vn, Phone: +84 90 428 8477.

The manufacturer's authorised representative in the EU is Springer Nature Customer Service Centre GmbH, Europaplatz 3, 69115 Heidelberg, Germany. If you have any concerns regarding our products, please contact ProductSafety@springernature.com

Printed and bound by CPI Group (UK) Ltd, Croydon, CR0 4YY
30/09/2026
02225471-0005